Outer
Continental Shelf
Frontier Technology

Proceedings of a Symposium
December 6, 1979
National Academy of Sciences

Conducted by the
Marine Board
Assembly of Engineering
National Research Council

NATIONAL ACADEMY OF SCIENCES
Washington, D.C.   1980

The National Research Council was established by the National Academy of Sciences in 1916 to associate the broad community of science and technology with the Academy's purposes of furthering knowledge and of advising the federal government. The Council operates in accordance with general policies determined by the Academy under the authority of its Congressional charter of 1863, which establishes the Academy as a private, non-profit, self-governing membership corporation. The Council has become the principal operating agency of both the Academy of Sciences and the National Academy of Engineering in the conduct of their services to the government, the public, and the scientific and engineering communities. It is administered jointly by both Academies and the Institute of Medicine. The Academy of Engineering and the Institute of Medicine were established in 1964 and 1970, respectively, under the charter of the Academy of Sciences.

This report represents work supported by Contract Number N00014-76-C-0309 between the Office of Naval Research and the National Academy of Sciences.

International Standard Book Number 0-309-03084-6

Library of Congress Catalog Card Number 80-82152

Available from:

Office of Publications
National Academy of Sciences
2101 Constitution Avenue, N.W.
Washington, D.C. 20418

Printed in the United States of America

FOREWORD

For the past decade the Marine Board of the National Research Council has been assessing, among many matters, the available technologies for developing the nation's offshore oil and gas resources. In the course of its work in this connection, the Marine Board organized and conducted a symposium on the various technologies relating to the exploration and production of energy resources in the Outer Continental Shelf (OCS). The symposium took place Thursday, December 6, 1979, at the National Academy of Sciences in Washington, D.C. This publication contains the papers presented on that occasion.

Co-chaired by two members of the Marine Board, Ronald Geer and James Rickard, the symposium had two objectives. The first was to enable some of the most knowledgeable experts from the industries engaged in offshore oil and gas operations in the harsh new frontier characterized by deep water and Arctic ice to provide their best evaluation of the technical capability and experience. This information will be immensely useful to the Marine Board's Committee on Assessment of Safety of Outer Continental Shelf Activities in its study of oil and gas drilling and production. The committee's study is described by its chairman, George Mechlin, on pages 13-15.

The second objective served to fulfill the Marine Board's continuing obligation to "...provide a forum to facilitate the exchange of information and data" on national ocean engineering issues, opportunities, and capabilities. In this regard, the Board considers that increased awareness of the offshore industry's technical capabilities and future directions would be most beneficial to government people and other concerned individuals. Thus, the Board invited those who are increasingly confronted with policy, regulatory, and managerial questions relating to the use and impact of technology on OCS frontier areas. This proceedings is a further effort to meet the second objective.

In introducing the symposium, Ron Geer raised several important questions that the speakers were asked to address:

- What are the limits of the current technologies in use in OCS oil and gas operations, technically and economically, and how can the technologies be extended to the frontier areas to best effect without increasing the environmental risks?

- What new drilling and production systems and equipment will be technically and economically feasible to advance the development of the new offshore tracts?

- When will the new technology and equipment be available to the industry?

Answers to these provocative questions came in the ten papers and general discussion that followed. Summarizing the day's activities, Jim Rickard noted that:

- Because of the uncertain economic potential of deep water offshore tracts, the industry has not drilled exploratory wells in water deeper than 4,876 feet.

- A fixed bed platform that is economically competitive with alternate types of platforms is not available in water deeper than 1,200 to 1,500 feet.

- Compliant platforms using guyed towers and tension leg platforms do not appear to be feasible in waters deeper than 2,500 feet.

- Technology does not now exist for installing subsea wells and associated facilities in waters deeper than 3,900 or 4,000 feet. Commercial discoveries have so far limited subsea operations to about 400 feet.

- Gravel islands and offshore platforms have not been fully developed for use in the Arctic Ocean off Alaska and Canada in waters deeper than 60 to 80 feet, although preliminary designs have been developed for building such structures in 200 feet of water.

- Existing equipment cannot be used to install large sized pipelines in waters deeper than 3,000 feet. The largest pipe now commercially available for deep water is 36 inches in diameter. While larger sizes are technically possible, the mills do not at present have a market for larger pipes.

- Floating production facilities cannot be utilized at the present time in water deeper than about 5,000 feet until more effective mooring and riser systems are developed.

- The capability exists to provide underwater inspection, maintenance, and repair services safely and efficiently in any water depth presently being considered for OCS frontier operations.

The speakers were uniformly optimistic about the industry's ability to advance the technology. The major constraints, in their view, are the high costs of deep water and Arctic operations and the uncertainties of the economic potential of energy resources in these areas.

It gives me great pleasure to express the appreciation of the Marine Board to the outstanding speakers who so willingly gave their time and effort to take part in the symposium.

Those who attended the conference are listed on pages 219-224.

Ben C. Gerwick
Chairman
Marine Board

University of California, Berkeley
March 14, 1980

vi

TABLE OF CONTENTS

INTRODUCTION

Ronald L. Geer

Jim Rickard and I are most pleased to be with you today as participants in the symposium and to have the opportunity of commenting on a subject that is of importance to us all--namely, the technologies available to develop the nation's oil and gas resources on the OCS.

The importance of the OCS and its potential for supplying a significant part of today's energy needs, and an increasingly significant part of tomorrow's needs, is unquestioned. The great uncertainty, however, is just how much of our future oil and gas supplies will come from the offshore areas, and, from a total national viewpoint, what is the best way to develop and utilize the OCS energy resource potential when balanced against all other alternatives.

We have assembled here today a group of recognized industry experts who will review for you some of the scientific and technological problems that confront the petroleum industry in its efforts to contribute to the developments of the OCS oil and gas resources.

We will discuss, without belaboring or attempting to quantify the tremendous efforts required, the status of offshore technology, particulary as it pertains to deep water and arctic frontier areas.

As I am sure you are aware, new technology and new applications of old technology are emerging at a rapid rate to meet the demands of the industry as it moves forward into deeper and more hostile waters and into the arctic frontier. The overall increase in activity in those areas has introduced some challenging technical problems.

Some of the questions raised by these problems are:

- What are the limits, both technical and economical, of current techniques, and how can they be extended profitably and without increasing any environmental risks?

- What new drilling and production systems and equipment will be technically and economically feasible to permit development of the new offshore areas?

● When will this new technology and equipment be available by and for the industry?

The significance of these questions is readily apparent when the United States offshore continental shelf is considered in terms of its petroleum potential and the nation's need for early development of this potential to help achieve energy self-sufficiency.

Before proceeding, I would like to provide a very brief overview of the importance of the offshore oil and gas resources to our nation's energy supply and also characterize for you the petroleum industry's present capability to explore, develop, and produce oil and gas resources in deep water and in arctic frontier areas.

Figure 1 shows the trend of total onshore and offshore domestic oil and gas production in terms of millions of barrels per day, with the gas expressed as BTU equivalent in barrels. As you will note, while the total domestic production has declined steadily, and is expected to continue to decline, offshore production has increased and is expected to increase further. Its share of the total is forecast to be approximately 27 percent in the 1990's as compared with some 19 percent in 1980 and about 15 percent in 1970. This forecast is based on the premise that the federal government will continue to offer about the same amount of acreage per year that has been proposed for the currently scheduled lease sales.

However, in order for the forecast to be met or, hopefully, exceeded, acreage that is considered by the oil industry to be the most promising must be made available.

The tremendous potential for domestic offshore production is indicated by Figure 2, which shows approximately 890 million acres or about 1.4 million square miles of submerged land from the shoreline out to a water depth of 8,000 feet, roughly the edge of the continental margin. Approximately 409 million acres (comprised of 342 million acres in the non-arctic and 67 million acres in the arctic-related environments) are within reach with today's technological capability.

Of this, only some 17 million acres or about 4 percent are currently under lease. I think it is quite clear that the offshore areas offer considerable potential for an increasingly important role in the nation's energy supply. It is critically important to our nation's future that all promising offshore areas be explored in an expeditious manner to determine the real potential for production.

Concern has been expressed recently by various groups that deep water acreage and arctic frontier areas should not be offered for lease because of the lack of adequate technology for producing in deep water. Over the last several years, the petroleum industry

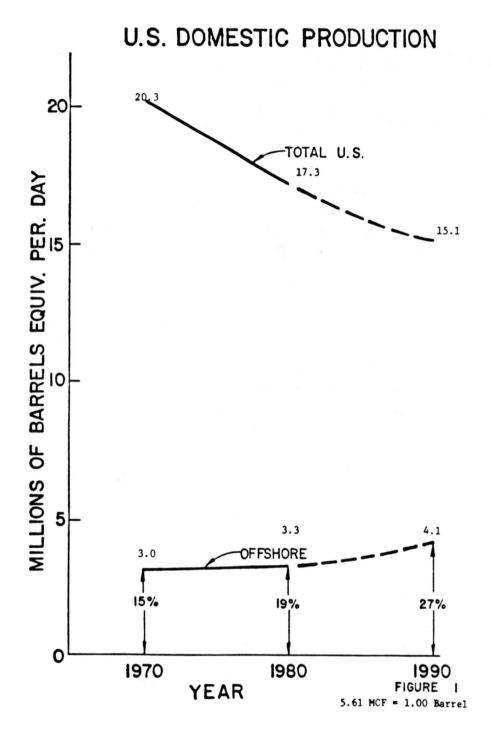

U.S. DOMESTIC PRODUCTION

FIGURE I

5.61 MCF = 1.00 Barrel

THE 890,000,000 ACRES
ON THE CONTINENTAL
MARGINS OF THE U.S.A.

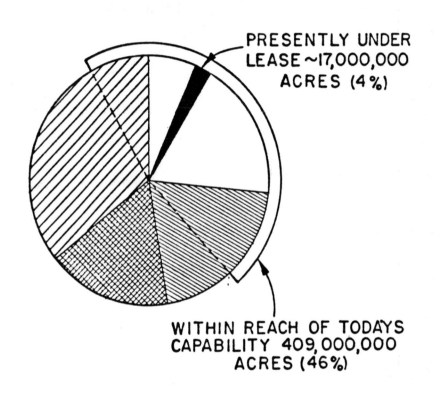

PRESENTLY UNDER
LEASE ~17,000,000
ACRES (4%)

WITHIN REACH OF TODAYS
CAPABILITY 409,000,000
ACRES (46%)

SHELF
SLOPE   ALASKA NORTH OF
          PENINSULA

SHELF
SLOPE   CONTERMINOUS U.S. +
          ALASKA PACIFIC

FIGURE 2

has spent tens of millions of dollars in designing equipment, developing analytical tools, performing model tests, and conducting full-scale experiments in the open ocean, as well as in the Arctic, in order to be prepared to drill and produce oil and gas resources in water depths out to 3,000 feet in the non-arctic areas and in some lesser water depths in the ice-laden arctic areas of the Alaskan North Slope.

Figures 3, 4 and 5 illustrate what is perceived to be industry's current capabilities. Figure 3 illustrates the industry's capability, which is based on results of work today. You will see and hear of some actual experiences in deepwater drilling. The figure shows that the capability for floating drilling as well as subsea completions exists for water depths up to and perhaps in excess of 6,000 feet. The record water depth to date is in some 4,876 feet off the coast of Canada.

Figures 3, 4, and 5 illustrate what is perceived to be floating drilling capability. However, the record to date is 620 feet off Brazil. Now, the reason that subsea wells are not being completed in greater water depths is not because of lack of technology but simply because of the lack of commercial discoveries.

Figure 4 shows that the capability presently exists for installing subsea facilities in water 3,000 feet deep. Again, lack of commercial discoveries has limited the actual installations to about 400 feet or less.

Figure 5 shows experience and capability with the fixed leg platform. The record water depth is presently 1,025 feet. The two dots on the curve represent the Hondo structure in more than 800 feet of water in the Santa Barbara Channel off California and the COGNAC structure in the Gulf of Mexico in 1,025 feet of water. It is expected that the fixed leg platform will be limited to a maximum water depth of 1,200 to 1,500 feet, primarily due to the cost of fabrication and certain installation constraints.

However, as you will see from Figure 6, new types of platforms such as guyed towers and tension leg platforms offer promise of extending platform capability significantly. The compliant structure shown here and the tension leg platform were test structures installed by Deep Oil Technology off Southern California. The guyed structure which Exxon installed in the Gulf of Mexico was also for test purposes to prove the technology and the theories and modeling behind them. They have since been removed.

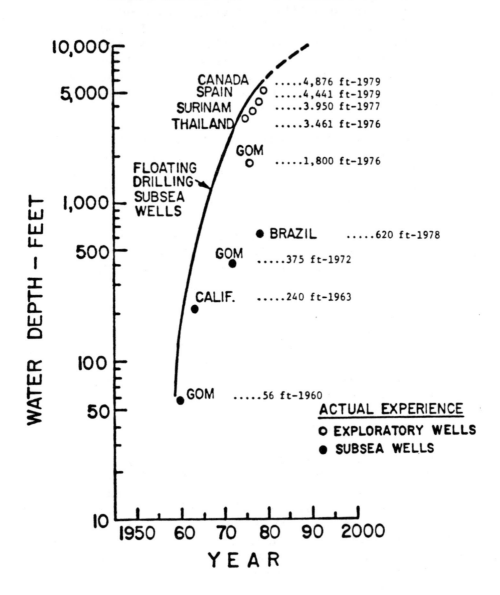

FLOATING DRILLING, SUBSEA WELL & FACILITY CAPABILITY PROJECTION

FIGURE 3

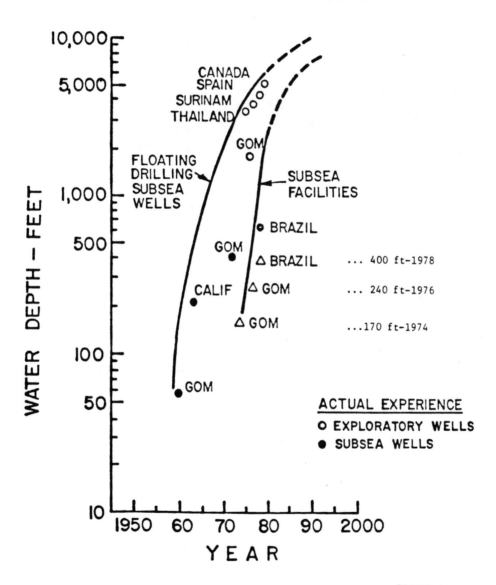

FLOATING DRILLING, SUBSEA WELL & FACILITY CAPABILITY PROJECTION

FIGURE 4

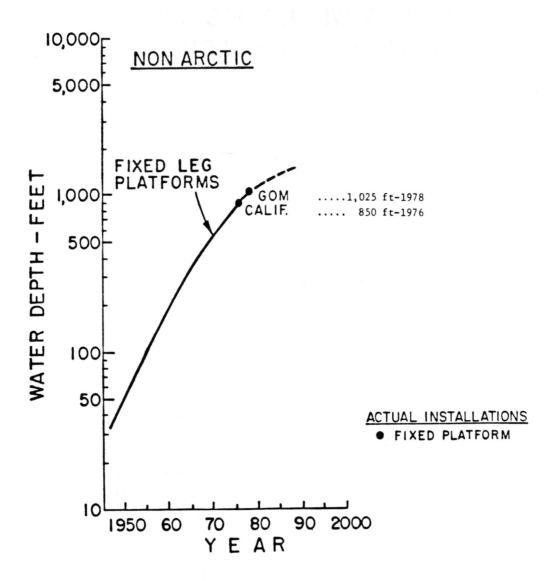

INDUSTRY PLATFORM
CAPABILITY PROJECTION

FIGURE 5

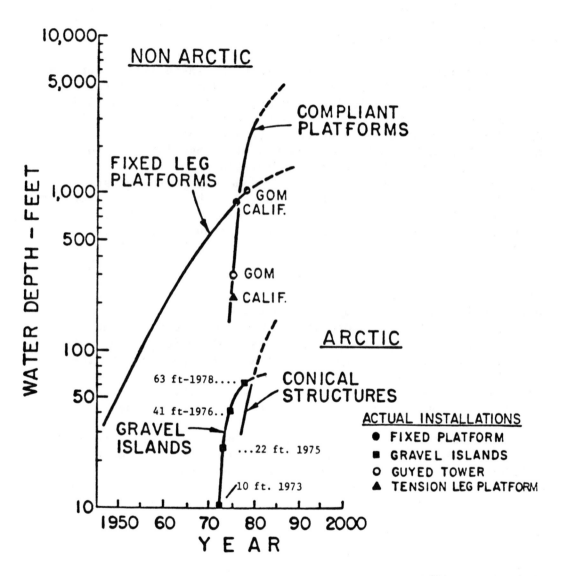

INDUSTRY PLATFORM
CAPABILITY PROJECTION

FIGURE 6

The figure also shows that gravel islands are expected to be applicable to about 80 feet of water. Experience has been gained in the Canadian arctic with this type of activity, which we expect to be almost directly applicable to the Beaufort Sea area in Alaska. It is expected that conical structures of a gravity type will be used in greater depths. The actual depth capability is yet to be determined.

Figure 7 shows the projected capability and experience for pipeline installation. In early 1980, a new depth record of 2,000 feet for such installations will be demonstrated in the Sicilian Straits in the Mediterranean Sea.

The net result of all this development and installation work is that the offshore petroleum industry is confident that the technology exists for exploring, developing, and producing oil and gas fields in water depths out to at least 3,000 feet off the coasts of the lower 48 United States and in certain ice-laden areas of the Alaskan arctic. This is not to say that, as you will hear later on today, all of the design details have been completed or all potential operational problems have been solved. Some uncertainties still exist and will continue to exist as mankind pushes back the frontiers of ignorance with regard to technological needs. There will be risks similar to those faced in the past, but the industry is confident that those risks can be handled in an orderly fashion as they have in the past.

The oil industry has an excellent record in extending the technology that is being used today around the world. About 85 percent of the technology used offshore in the world has been developed by U.S. companies. The only way that the technology for deeper waters and arctic areas will be extended is by actual experience in these areas. If the opportunities to design, develop, install, and operate systems where a real need exists are not made available to the industry, then the technology will not be developed nor demonstrated.

Now, before we proceed to detailed discussions of this technology, I would like to introduce George Mechlin, who will comment briefly on the Marine Board's Committee on Assessment of Safety of OCS Activities.

Mr. Geer is a member of the National Academy of Engineering as well as of the Marine Board. In June 1979, he was appointed vice-chairman of the Marine Board. For the past fifteen years, Mr. Geer has been associated with the Shell Oil Company, Houston, Texas, involved with Shell's pioneering offshore deep water drilling and production research development activities.

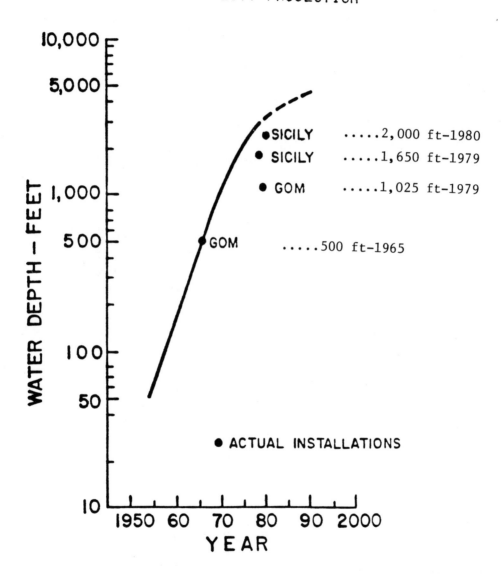

INDUSTRY PIPELINE
CAPABILITY PROJECTION

SICILY 2,000 ft-1980

SICILY 1,650 ft-1979

GOM 1,025 ft-1979

GOM 500 ft-1965

● ACTUAL INSTALLATIONS

FIGURE 7

COMMITTEE ON ASSESSMENT OF SAFETY OF
OUTER CONTINENTAL SHELF ACTIVITIES

George F. Mechlin

This symposium is one of a number of activities that are intended to provide support for a study being undertaken by the Committee on Assessment of Safety of Outer Continental Shelf Activities. The committee operates under the auspices of the Marine Board of the Assembly of Engineering. The objectives of the committee's study are twofold: To provide a review and assessment of safety regulations and technology supporting oil and gas operations in the OCS, and to develop the methodology for technology and regulation assessment in future studies.

The origin of the study can be found in the Outer Continental Shelf Lands Act Amendments of 1978. This legislation, in Section 21(a), directs the Department of the Interior in which the U.S. Geological survey resides, and the Department of Transportation, in which the U.S. Coast Guard resides, to "...commence a joint study of the adequacy of existing safety and health regulations, and of the technology, equipment, and techniques...." This committee will provide the Geological Survey and the Coast Guard with a technical base and analysis for completing the 21(a) study.

The committee's work will not encompass the entire scope of the 21(a) mandate, but rather will address the technologies and regulations associated with the safety of OCS oil and gas drilling and production. In addition, the committee will develop criteria for judging the adequacy of personnel and environmental safety and a methodology for applying the criteria to future activities of this nature. This review and study will result in a final report to the U.S. Geological Survey.

Our definition of safety is protection of life and limb and the protection of the marine environment. We do not intend to comment on any specific industry but to synthesize the general capabilities of industry and provide that as the basis for the study.

We will limit the regulations to be reviewed to those that are in effect or proposed to 1 January, 1980. We are going to exclude the transportation systems. However, we will include pipelines. Futhermore, we will not attempt to delve into the issues of post-spill effects of oil on the environment or diving safety. These have been the subject of a number of studies, and we think there is very little that we can add to this very difficult technical subject at the present time.

Finally, we will not comment on any proposed organizational changes in the government nor on the effectiveness of any existing unit or agency of the government.

The members of the Committee are:

Dr. George F. Mechlin (Chairman)
Westinghouse Electric Corporation

RADM J. Edward Snyder, Jr.
(Vice Chairman)
U.S. Navy (Retired)

Mr. Willis C. Barnes
ORI, Incorporated

Dr. Michael E. Bender
Virginia Institute of Marine Science

Mr. H. Ray Brannon, Jr.
Exxon Production Research

Ms. Sarah Chasis
Natural Resources Defense Council

Mr. Douglas Foy
Conservation Law Foundation

Mr. Gary Lynn Kott
Global Marine Drilling Company

Mr. William Linder
Petro-Marine Consultants

Mr. Bramlette McClelland
McClelland Engineers Incorporated

Dr. John Moroney
Tulane University

Ms. Hyla S. Napadensky
IIT Research Institute

Mr. Myron H. Nordquist
Nossaman, Krueger & Marsh

Mr. O. J. Shirley
Shell Oil Company

Mr. Phillip S. Sizer
Otis Engineering Corporation

Dr. Lawrence R. Zeitlin
Lakeview Research Incorporated

As is customary in the organization of the Assembly of Engineering studies, members of the committee have been chosen to represent many expertises and fields of endeavor. Of the 16 members, seven come from some sector of industry, four come from some form of academic organization; two are associated with public policy organizations, and one is retired from the U.S. Navy. Five of the members of the committee are under 40 years of age, which may be of some significance. We are supported by a very capable staff organized by the Marine Board.

Dr. Mechlin, a member of the National Academy of Engineering, was chairman of the Marine Board from 1976 to mid 1979. As a member of the Marine Board, he also chaired its Panel on Operational Safety of Offshore Resource Development (1972) and the Review Committee on Safety of Outer Continental Shelf Petroleum Operations which produced four reports (1974-1975).

Dr. Mechlin is Vice President, Research, and General Manager, Research Laboratories, Westinghouse Electric Corporation. He has spent most of his business career working in advanced technology areas. He holds masters and doctors degrees in physics from the University of Pittsburgh where he was a graduate in 1944 with a Bachelor of Science degree. Dr. Mechlin is a member of a number of professional societies and is the recipient of the U.S. Navy Meritorious Public Service Award, the Westinghouse Order of Merit, and the John J. Montgomery Award.

EXPLORATORY DRILLING SYSTEMS

J. C. Albers

Introduction

My talk to you today concerns exploratory drilling systems in the
context of present-day state of the art. We are today drilling in
ever deep waters. This year, drilling has approached 5,000 feet of
water. The unmistakable trend is to probe even deeper waters in the
near future in the search for hydrocarbon reserves. As we are aware,
the search for hydrocarbons, like any activity, is not without risk.
In my talk, I will briefly review current trends and industry
capabilities and discuss in greater detail environment and how risks
have been overcome with safety built into the deepwater drilling
programs.

Trends

First, a brief description of industry trends and where we stand
today. In the 20 year period from 1952 to 1972, drilling offshore
progressed from 100 to 1,300 feet of water. Over the last eight
years, the drilling industry has increased its actual drilling
performance to almost 5,000 feet of water in 1979. This capability
has generally preceded actual experience by some two to four years.
Water depth capabilities have not been the only environmental
challenge requiring extension. The drilling industry has also had to
extend its capabilities concurrently to work in wind and waves, ice,
and currents and tides.

The drilling industry currently has a fleet of 13 vessels, 11
drill ships and two semisubmersibles capable of drilling in 3,000
feet or more, of water. Ten of the drill ships and one semi-
submersible are equipped with dynamic positioning (DP). Four of the
deep water drilling vessels are presently capable of drilling in
6,000 feet of water. Three are drill ships, and one is a semi-
submersible. The technology for extending the drilling capability of
certain of these vessels from 6,000 feet to 8,000 feet is available
today.

A total of 17 wells have been drilled in water depths of 3,000 feet, or greater, as of this date. One well was drilled in 1976 and one in 1977. In 1978, six wells were drilled. In 1979, nine wells were drilled in over 3,000 feet of water. Five of these wells were drilled off West Australia, three were drilled off East Canada where the Discoverer Seven Seas set the present record in 4,876 feet of water, and one well was drilled off the coast of Spain. No wells, to date, have been drilled in U.S. waters over 3,000 feet deep. Drilling has progressed over the last 28 years offshore from shallow coastal waters to the outer continental slope and has most recently probed the continental rise. It can be expected that in the 1980's, there will be more drilling on the continental rise, and that drilling will have been accomplished at 8,000 feet.

Technical Achievements

The ability to extend exploratory drilling capabilities to 6,000 feet or more today can be credited to four major areas of technical achievement. The first area is mooring. Dynamic stationing utilizing thrust provided by propulsion units has replaced mooring lines and anchors to maintain the drilling vessel over the wellhead. Dynamic stationing systems consist of four elements:

- Position references

- Computation center

- Thrusters

- Power Plant

The SEDCO 445 was the first dynamic stationed drill ship, going into service in 1971. From 1972 to 1976, this vessel drilled over 200,000 feet of hole on DP in an average water depth of 1,560 feet. During this period of time the downtime due to DP repair was 0.5 percent.

Re-entry is the second major technical achievement. Guidelines as used in shallow water drilling operations to guide the blow-out preventer (BOP) and other subsea equipment to the wellhead are not capable of supporting deep water operations. This has necessitated the development of drilling techniques utilizing television and sonar systems instead of guidelines. One major company utilizes beacon offset, outside BOP stack TV's, and through-the-riser TV, or bombshell TV, and combination TV/sonar through-the-drill pipe re-entry techniques.

Over 500 re-entries of BOP stacks, lower marine riser packages or bits and casing strings have been made.

Electrohydraulic BOP control has replaced the more conventional hydraulic control in deep water drilling, thus making BOP control the third major technical achievement. This change was promoted by the fact that the hydraulic system has an excessive signal transit time in deep water. For example, in 6,000 feet of water the hydraulic signal takes forty seconds to reach the BOP control pod on the preventer, whereas, the electrical system takes less than one second. The industry is already using a second generation electrohydraulic BOP control system utilizing multiplexing technology to permit BOP operation of 100 or more functions over a single pair of wires.

The fourth major technical achievement focuses on the riser. In shallow water operations, the application of tension at the top of the riser has been sufficient to hold stresses within tolerable limits while supporting the riser, the mud column, the environmental loads and rig offset. As the water depths increased, the tensioner numbers and the substructure requirements have dictated a need for other tensioning means. Buoyancy material has been developed to reduce the riser weight. The buoyancy material--a syntactic foam matrix with micro balloons and a fiber glass outside shell--has required improved manufacturing methods to insure reliability in deep water. The fiber glass outer shell provides a rugged outside cover to withstand routine handling abuse. Tests and experience have demonstrated that this material can be used for extended periods of time at 5,000 feet submergence with a loss of about two percent buoyancy.

The second achievement with respect to riser improvements has been in the riser coupling design. Past designs utilized a coupling that was approximately one-half the strength of the pipe tube. Current deep water designs are equal in strength to the pipe. Marine drilling risers have been used in 4,876 feet of water, in areas with a four knot current, and in areas with 15 feet average waves and with occasional 50 feet waves. Computer analyses indicate the present new riser system can be used in 8,000 feet of water with the requisite mud weights, rig offsets and environmental loads experienced in deep water.

Overcoming Risk With Safety

The following are descriptions of a few examples of overcoming risk with safety built into the deep water drilling program.

Riser Recoil

In operating a dynamically positioned drilling vessel, the possibility always exists that the vessel will undergo a "driveoff" in which the vessel moves away from its location over the well. The driveoff could be caused by excessive environmental forces, a computer malfunction or human error. In order to protect both the well and the rig during a driveoff, there are provisions to disconnect the upper BOP package and riser from the lower BOP stack by initiating a programmed "emergency disconnect sequence" which closes certain valves on the lower stack and unlocks the riser connector.

In deep water, a surface tension of 600,000 lbs. or more may be pulled on a buoyant marine riser filled with drilling mud. If this tension were to be maintained during the actual disconnect, the tension would pull the upper stack clear of the lower stack and continue pulling and accelerating the suspended riser system until the terminal velocity of the systems in water was reached. The system would keep traveling upward at a constant velocity until the elements of the slip joint slam together and the energy of the system is transferred by impact to the structure of the drilling vessel. This could cause extensive damage to the vessel as well as buckling failure of the marine riser.

To prevent such an occurrence, a number of deep water vessels have been equipped with riser recoil preventer systems. These systems enable the creation of a closed hydraulic system on the tensioners at the time of disconnect. With an essentially closed system, an over-tension on the suspended riser system can exist which will pull the upper stack up and away from the lower stack. This allows the tensioners to stroke out, increasing the volume in the closed system and decreasing the pressure to the point where the suspended system comes into equilibrium and slowly stops. Normal lift-off is in the range of 10 to 15 feet.

Recovering a Dropped BOP Stack

While retrieving a BOP stack offshore Australia in 3,150 feet of water, the lower stack dropped 1,800 feet to the bottom due to a mechanical malfunction in the riser connector. Since all BOP running and retrieving operations are conducted with a 100 foot vessel offset from the well, the stack did not hit the wellhead but landed harmlessly on its side into a very soft bottom. Several attempts were made to fish for the stack without success. A second onboard BOP was then put into operation to resume drilling operations. While drilling operations were on-schedule with the second BOP, efforts were initiated to come up with a successful recovery technique. Several possibilities were investigated, including attaching air bags to the BOP with a manned submersible.

Two recovery tools were finally selected for the task. The primary tool was an "internal" tool, and the backup tool was an "external" tool, both run on drill pipe and operated from the ship. The internal tool consisted of a counterweighted assembly, horizontally mounted on a drill pipe sub having an expandable locking ring on the nose, a jet assembly, and a TV for vision. The horizontal mounting was pivoted and could be changed in angle by some hydraulic locks. The tool was designed to be inserted into the horizontal riser mandrel, lock, and begin retrieval. The external tool was essentially a grapple designed to fit over the horizontal BOP and hydraulically lock onto the stack at its most structurally-sound lift points.

The internal tool, being the primary tool, was used first during an interruption in the drilling of the well. After lowering the tool on the drill pipe close to the sea floor, the stack was sighted, latched-on to, and stack pulling initiated within a 30 minute time frame. Overpull was only 10 to 15k. The BOP stack was successfully recovered to the ship within four hours with only minor damage. The BOP stack was placed on its test stump and tested successfully to 10,000 PSI, its rate working pressure.

Design of a Reliable and Safe BOP Control System

To illustrate the measures taken to develop reliable and safe offshore systems, I'd like to review one program directed in this regard on the multiplex (MUX) BOP control system for deep water drilling. In order to come up with the best state of the art system to do the job in a reliable and safe manner, the following steps were taken in design and construction.

Design. The control system is 100 percent redundant; that is, there are two separate electronic, electrical and hydraulic systems subsea to provide control. Should one system go down for any reason, the other system is fully capable of handling all operating functions. The fact that one entire system fails has no bearing on the functioning ability of the other system. Within each of the control cables for each of the two fully redundant MUX control systems are a number of separate individually paired wires which served as a hardwire electrical backup to the multiplex systems. The hardwire systems serve only the most critical BOP stack functions.

Supplementing the two MUX systems and the two hardwire backup systems is a two-way acoustical control package, again only to the more critical BOP functions. Thus, there are five circuits to operate critical functions!

In addition to the five operating systems, there are two other systems which can be brought into play during well control procedures. The first is called "auto-shear." This is a mechanical "mousetrap" which causes the shear rams to close whenever the riser package is disconnected from the lower BOP stack. This function is armed and disarmed through the control system. While drilling, it is operating practice to have it in the armed condition, ready to operate. In the event of a control system failure, it remains in the armed condition. Complementing the auto-shear is an emergency recovery system which provides for control of limited functions by stabbing a drill pipe probe into one of two funnels on the lower marine riser package. Thus, in effect there are eight different ways of controlling critical functions on a deepwater BOP stack on the sea floor without the need for manned subsea intervention. These systems are dependable and safe, and relied upon by the operating people responsible for them. They do require operating technical expertise above and beyond conventional operations.

Construction. Several measures were taken during construction and assembly of these new MUX BOP control systems to ensure that they would be reliable and safe. The first effort in this regard was to test the electronic and hydraulic portions of the control system in a hydrostatic test tank to 6,000 feet of salt water submergence. All functions were cycled repeatedly in the tank to ensure their ability to operate in 6,000 feet of water.

The second effort to ensure reliability and safety of operation was to subject the surface and subsurface electronics to an adequate "burn-in" test. All electronics have a certain degree of "infant mortality" depending on the level of specification of these components. All electronic components were specified to the highest available industrial level of military specification reliability. After integrating these components into the system, each BOP function, whether an open-block-close, an open-close, an extend or retract or a readback--was tested, after the above hydrostatic test, to a minimum of 20,000 cycles each. If there was more than one wrong execution, or one unexecuted command per 10,000 cycles as recorded by the computer, the test was started again until it ran through a new set of 20,000 cycles within this specification.

Each of the tested systems passed with flying colors, in fact, we became concerned we might over-test and have the equipment in an "old-age mortality" area before it ever got to the field. This concern has not proved founded.

Conclusions

Exploration drilling technology is presently capable of providing reliable and safe drilling in 6,000 feet of water. We have already demonstated this capability in almost 5,000 feet of water. Existing exploratory drilling vessels should be capable of extending their capability to 8,000 feet of water.

Drilling technology and capabilities generally run some eight years ahead of production technology, with good justification. There is need for drilling, but the reserves in deep water necessary to justify advances in production technology have not yet been discovered. Drilling technology is providing a proving ground for future production technology from totally subsea or stationary, but not fixed, surface platforms. Exploratory drilling in deep water is an expensive proposition. Rates are in the $50,000/day+ range. Capital costs are 50 to 75 million dollars and people are more important than ever to the effort. The industry's current capabilities, while the result of the successful development of technical advances, took detailed engineering and operating expertise, a responsible industry management and long-term R&D efforts.

Mr. Albers is Systems Engineering Manager for the Drilling Division of SEDCO, Incorporated in Dallas, Texas. In this capacity, he is responsible for engineering development of SEDCO's deepwater drilling systems.

Mr. Albers joined SEDCO in 1975 after nearly seventeen years with Shell Oil Company. His education includes undergraduate studies at St. John's University (Minnesota); B.S., Mining Engineering, (Petroleum Option) University of North Dakota, 1958; Graduate studies at the University of Alaska. He is a member of several technical and professional societies.

DEEPWATER DRILLING AND PRODUCTION PLATFORMS IN NON-ARCTIC AREAS

F. P. Dunn

(Mr. Dunn opened his presentation by showing a motion picture of the fabrication and installation of the COGNAC platform located in Mississippi Canyon Block 194 in the Gulf of Mexico. The field lies some 15 miles south of the mouth of the Mississippi River on four blocks covering approximately 22,500 acres. The platform is operated by the Shell Oil Company for a group of 15 companies.

COGNAC is the world's deepest water platform (1,025 feet) and also the world's heaviest steel platform--59,000 tons. It is also the world's first three-part platform. The base section of the platform is 380 feet by 400 feet in plan, and 175 feet high. It weighs 14,000 tons. The midsection of the platform is 282 feet by 310 feet in plan and 315 feet high. It weighs 8,500 tons. The 11,000 ton top section, which is 207 feet by 254 feet in plan and 530 feet high, will support a 2,500 ton deck with two complete drilling rigs. There are 24 piles driven, each 84 inches in diameter and 615 feet in length, driven through the base, 450 feet into the sea bottom. The COGNAC jacket weighed 33,000 tons. For comparison, the jacket for the Exxon Hondo platform, located in the Santa Barbara Channel off the coast of California in 850 feet of water, weighs 12,000 tons.

The platform is designed for 62 wells to be drilled by two rigs. Production from the platform has commenced, and is being produced initially via pipeline to Shell's East Bay Central facilities located onshore south of Venice, Louisiana. Initial production rates were about 5,000 barrels of oil per day which will gradually increase to 50,000 barrels per day by late 1981, when drilling will be completed. Ultimate production will be about 100 million barrels of oil and 500 billion cubic feet of gas. The productive life of the project is expected to be some 15 to 20 years. The total investment is expected to be between 750 to 800 million dollars.)

I would like to talk for a few minutes about the industry's deep water capabilities. Today there are five platforms installed in over 500 feet of water around the world. Four more are being fabricated, and at least five additional are being designed. Union Oil, for example, now is fabricating a platform in one piece from 935 feet of water to be installed in the Gulf of Mexico. British Petroleum has just started building their Magnus platform. The jacket alone weighs 40,000 tons. Exxon is actively contemplating installing a guyed tower, about which I will say more in a few minutes, in about 1,200 feet of water in the Gulf of Mexico.

The curves in Figure 1 indicate that on a cost basis alone, we would tend to switch to a guyed tower in about 800 feet of water. However, the transition is more likely in 900 to 1,100 feet of water, depending upon specific circumstances, such as location. This is a fairly representative cost curve for the Gulf of Mexico, for a relatively small platform.

There are various bases for choosing the platform configuration. The important two are cost and reliability, or one's perception of reliability.

The COGNAC platform was built in three pieces because there were no fabrication facilities for one piece of this size at the time.

Union is now building a platform in one piece for installation in 935 feet of water. A one piece platform will not take waves much over 30 to 35 feet, possibly even less depending upon the configuration, until at least four or eight piles are installed. Thus, there is a period of a week or so in which such a platform is vulnerable unless it is installed, say in May or early June when the probability of getting even 30 or 35 foot waves is remote.

There is another type platform, a self-floater which does not require a launch barge. Figure 2 is an example of such a type. British Petroleum's Magnus platform is a similar type. No barge now exists which is capable of carrying a platform weighing that much. The largest barge can carry a platform of about 30,000 tons.

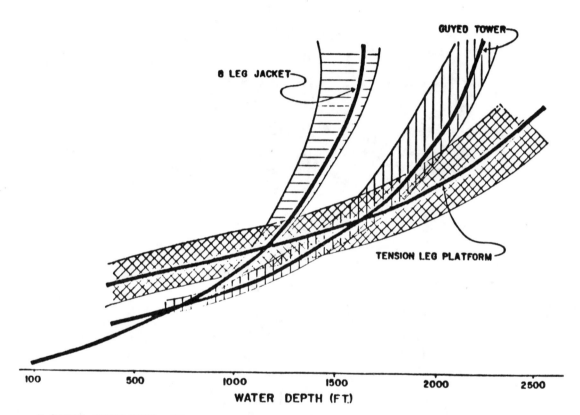

WATER DEPTH (FT.)

✻ DESIGN, FABRICATION AND INSTALLATION COST, EXCLUDING TOP SIDE EQUIPMENT AND FACILITIES.

FIGURE 1

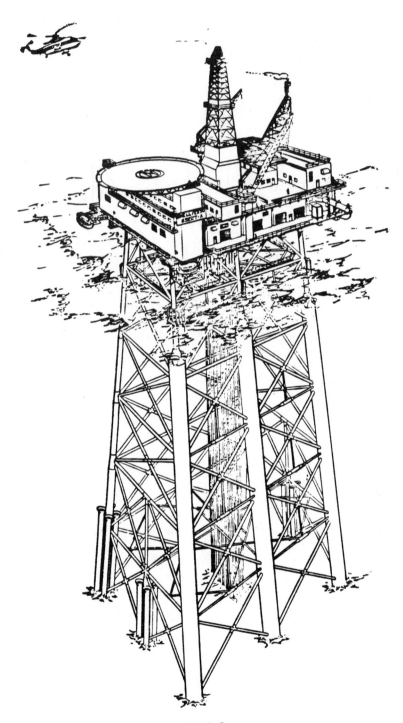

FIGURE 2

The disadvantage of a self-floater is that a graving dock somewhere may be required for fabrication and, of course, extra steel is
required for the flotation.

Figure 3 shows a concrete gravity platform. There are none in
the U.S. waters except in fairly shallow water. Platforms such as
illustrated in Figure 3 are huge. This one weighs about 250,000
tons. There are now 14 of these monstrous platforms in place. They
have storage capacity for up to 500,000 barrels of oil. Another
advantage is that, in principle, they can be taken out, installed and
are almost immediately ready for drilling operations. Topside
facilities can be carried on them, which is indeed a big advantage,
if you do your scheduling properly, no mean feat. If there is no
pipeline, you can store a good deal of oil and then ship it from an
offshore loading terminal.

The steel gravity platform shown in Figure 4 has the same
advantages as the concrete gravity platform and the same disadvan-
tages--one of which is cost. There are a few in fairly shallow water
offshore western Africa, and there is now one being fabricated
similar to Figure 4 for the North Sea, the first large steel gravity
platform to be installed there, in about 500 feet of water. The
platform has storage capability, and it also requires a fairly firm
foundation similar to the concrete gravity platform.

Compliant structures are illustrated in Figure 5. These structures will be used for the most part in water depths over 1,000 feet,
though there is one, a tension leg platform, which will be installed
in the North Sea in about 550 feet of water.

The upper left illustration in Figure 5 is an articulated
column. Thus far, there is no production or drilling platform of
this type. There are a few loading platforms of this type. One is
located off Brazil, and there are a couple located in the North Sea.
They are not practical for use as drilling and production platforms.

In the lower left of Figure 5 is a semisubmersible. These can
be converted into production platforms. Their disadvantage is that
they are quite sensitive to topside weight. This is true, of course,
of all of these buoyant type structures, and it is difficult to add
substantial production facilities to them, unless, of course, the
structures are very large.

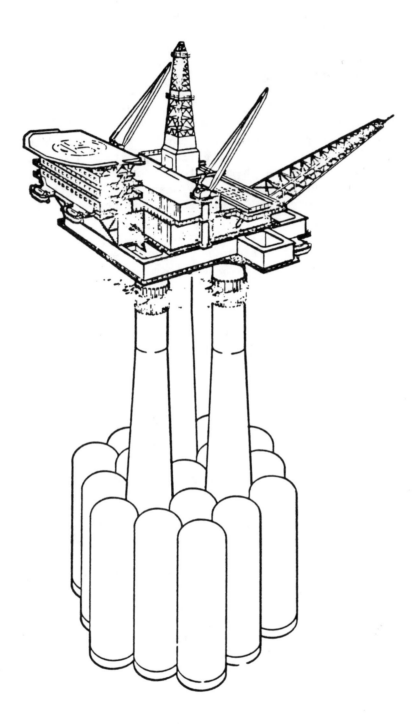

Figure 3 - Typical Concrete Gravity Platform

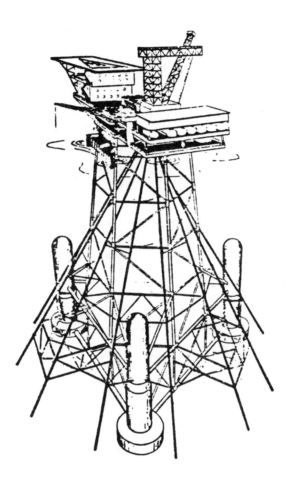

Figure 4 - Steel Gravity Platform

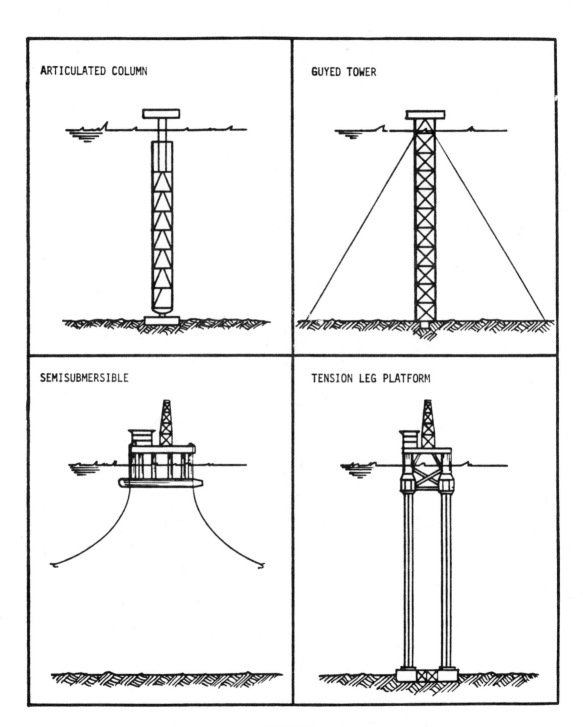

FIGURE 5

In the upper right of the figure is the guyed tower. I will talk a little more about this in a minute. Finally, the tension leg platform platform is shown. It will probably be utilized in water depths beyond 1,000 feet--or shallower. The choice depends on costs and the individual engineer's perception--call them prejudices-- whatever.

Figure 6 is a picture of the guyed tower. It is a relatively simple structure that is not difficult to fabricate and is not heavy. Its movement is altogether tolerable, and it is held in place with 12, 16 or 24 cables, which could be wire rope, synthetic material, chains, or combinations thereof. A lot of analytical work has been done on these, and there is no doubt but that they will work.

There are two ways of anchoring the base. One is by pure gravity. Implant the base in the ground, using a large spud can. Long term settlement, obviously could be a problem. This can be taken care of by preloading. The other means of anchoring the guyed tower is to install a piled foundation. I am in favor of such a foundation because it reduces settlement uncertainties.

There are several versions of the tension leg platform (Figure 7). In one type, the wells are subsea; there is manifold on bottom, and one to four risers are brought up to the surface. Such platforms are quite sensitive to topside load, understandably, and they also need buoyancy for tensioning the anchor line and the risers.

The tensioning anchors can be either wire rope or regular pipe. Several designs are available with 18- or 24-inch pipe as the tensioning devices. The pipe has the obvious advantage of more corrosion resistance than wire rope. Moreover, the fatigue problem with wire rope, as several of you know, can be considerable.

The price tag on a North Sea type tension leg platform is over $500 million. It is this high because of the need for large production facilities for about 150,000 or 200,000 barrels a day. Keep in mind that in the North Sea, if production is only 5,000 to 15,000 barrels a day as it usually is in the Gulf of Mexico, the operation is uneconomic.

There is another version of the tension leg platform wherein the wells are brought to the surface through the tensioning devices. The tensioning devices do double duty. They act as well conductors and serve as the anchors. The wellheads are on the surface.

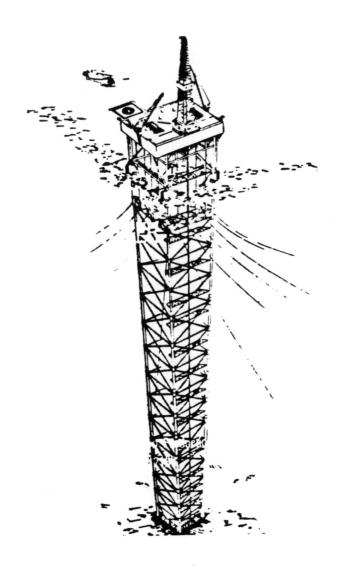

Figure 6 - Guyed Tower Concept

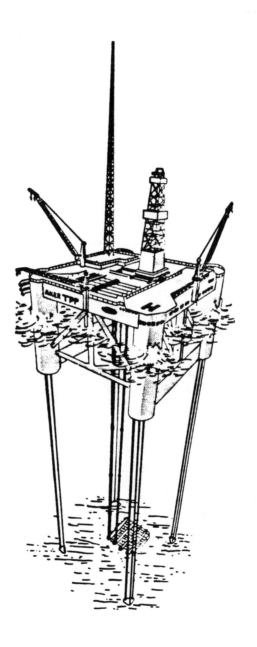

FIGURE 7 - TENSION LEG PLATFORM

Figure 8 is somewhat of a hybrid. The wells are all brought to
the surface, so the platform is similar in principle, at least, to a
regular fixed platform. Everything can be done from the surface.
The tensioning devices are separate, however. The base weighs about
10,000 tons. The superstructure can weigh anywhere from 15,000 to
25,000 tons. This is a design for California or the Gulf of Mexico.

Figure 9 is a list of areas requiring further work. In my
opinion, the most important need today is not analysis. Field
measurements are needed to correlate and to calibrate our analysis.
We do not have enough, and we must spend the money and time to obtain
such measurements.

I am very optimistic about our own program of instrumentation
for COGNAC. Also, we need more information on response of structures
to earthquakes. There are not very many structures out there, and
thank God we do not have too many earthquakes, but from an engineer's
standpoint, we need earthquake measurements from some platforms.

Thus, upgrading and calibration of our analytical tools follows
upon obtaining field measurements. We need the field measurements
first. And, of course, we need more model studies and laboratory
testing. We must have a better definition of pile capacity. We have
been installing piles now for 30 years, and we still do not know as
much as we should. Keep in mind that we are talking about pile
loadings of 10,000 and 12,000 tons. They are not like these 100,
200-ton piles that are holding up, say, this building. They are huge
piles, as much as 96 inches in diameter. It is a little shaky to
extrapolate from 24-inch diameter piles to 96-inch diameter piles.
We just need more information.

The last item is, in my opinion, the most important. We need
engineers to take care of the important details that managers sometimes
overlook. We need engineers--engineers who think engineering all the
time. We also need the opportunity in the form of deep water and
Arctic leases; and, of course, we need successful geophysicists and
exploration geologists.

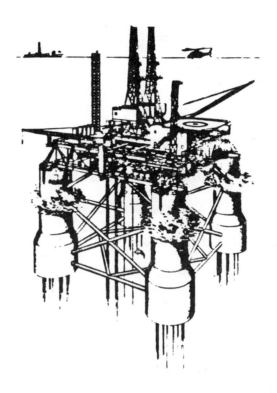

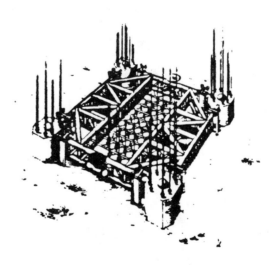

FIGURE 8 - TENSION WELL PLATFORM

AREAS REQUIRING FURTHER WORK

I. FIELD MEASUREMENTS

    A. STATIC/DYNAMIC RESPONSE

        1. NEAR-DESIGN LOADING CONDITIONS

        2. FATIGUE LOADING CONDITIONS - LOCAL MEMBER RESPONSE

    B. ENVIRONMENTAL DATA - WAVES, CURRENT, EARTHQUAKES

        1. OCCURRENCE FREQUENCY

        2. FORCE MEASUREMENTS

    C. FOUNDATION PERFORMANCE - PILED, GRAVITY

II. UPGRADING, CALIBRATION OF ANALYTICAL TOOLS

    A. MOTION ANALYSES - STRESS ANALYSIS FOR BUOYANT
        STRUCTURES, RISERS

    B. DYNAMIC ANALYSES

    C. FATIGUE ANALYSES

    D. RELIABILITY ANALYSES - ULTIMATE STRENGTH

        1. GOAL: BALANCED DESIGN

III. LABORATORY TESTING - MODEL STUDIES

    A. GEOTECHNICAL TESTING - FIELD, LABORATORY

    B. MODEL TESTS - WAVE TANKS

IV. ENGINEERING TALENT

FIGURE 9

In summary, there is no technological roadblock. We can now design and build these platforms for 2,000 to 3,000 feet of water. We need to do a lot more work, of course, for water that is deeper than 3,000 feet. I personally do not see anything that is going to hold us back other than money and opportunity. The previous speaker, Jim Albers, said that exploration drilling is eight years ahead of us. Well, we will catch up with them. Thank you.

Mr. Dunn is Manager, Civil Engineering Department of Shell Oil Company in Houston, Texas. A native of Springfield, Ohio, he holds a Bachelor of Arts degree from Xavier University, Cincinnati, and bachelor and masters degrees in civil engineering from Ohio State University. Prior to his present position, he was head of the Offshore Construction and Design Group for ten years. This group designs all Shell Oil platforms.

ARCTIC PLATFORMS

Hans O. Jahns

Introduction

In Arctic waters, the presence of sea ice poses a new challenge to
the designer of offshore structures for oil field development. Water
depth is also a concern, but present capabilities are on a very
different scale compared to the ice-free oceans. In sea ice environments, the term "deep water" may be appropriate in water depths
exceeding 100, or 60, or even 40 feet, depending on the severity of
the ice cover and the type of offshore structure in question. In
contrast to the previous speaker, I will be limiting my discussion to
OCS areas less than about 650 feet--about 200m--deep.

Figure 1 shows the shelf areas around Alaska inside the 200m
water depth contour. Not all of these areas experience sea ice. But

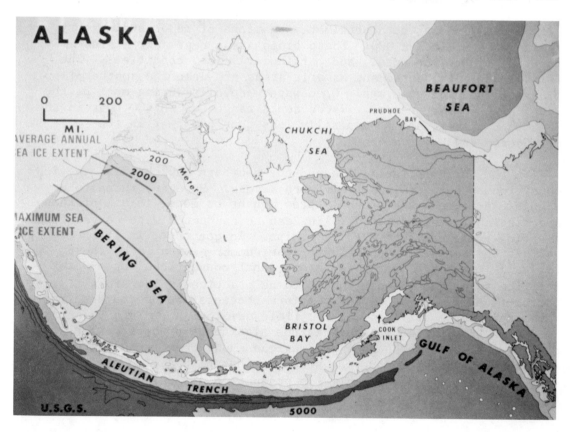

Figure 1

I have interpreted the term "Arctic" rather loosely: I plan to address platform concepts for all those Alaskan OCS areas where sea ice is a design consideration. In addition to the true Arctic areas in the Chukchi and the Beaufort Sea, this includes the sub-Arctic OCS areas in the Bearing Sea and in Cook Inlet. The heavy line indicates the approximate southern limit of sea ice at its maximum extent in bad ice years. The dashed line shows the maximum extent of the sea ice in an average ice year.

The industry has been operating successfully in two sea ice areas--in Cook Inlet in the south, and in the Beaufort Sea in the north, both off Alaska and in Mackenzie Bay on the Canadian side. In the first part of my discussion, I want to describe briefly the platform types that have been used, and the technology that has been developed for safe operations in these existing lease areas. The second part of my talk will address platform concepts that would be applicable, in our opinion, to those future lease areas in the Arctic Ocean and the Bering Sea, that are currently included in the Department of Interior's five-year leasing plan.

Figure 2 shows the approximate locations of major sedimentary basins on the Alaskan OCS. These basin areas represent a wide spectrum of environmental, and particularly ice, conditions. The Bristol Bay, St. George and Navarin areas are near the southern limit of annual sea ice coverage. Open water prevails during most of the year, nine months or more. Water depth ranges up to 300 feet in the Bristol Bay area, and is generally less than 650 feet in the St. George and Navarin basins.

In the northern Bering Sea and the southern Chukchi Sea, the ice conditions become progressively more severe. However, there is a significant open-water season, three months or more, in all of these areas so that exploration drilling can be conducted with conventional floating equipment such as drill ships. In contrast, in the Beaufort Sea off the North Slope of Alaska, the summer season is so short, typically one month in exposed areas, that conventional open-water drilling techniques do not appear to be practical. Therefore, bottom-founded, ice-resistant platform concepts are needed here for both exploration drilling and oil field development. This is an important distinction which makes the Alaskan Beaufort Sea, and perhaps the northern Chukchi Sea, unique among frontier areas. In all other areas, including the southern Chukchi Sea, ice-resistant structures will be needed only for production, if and when commercial discoveries are made.

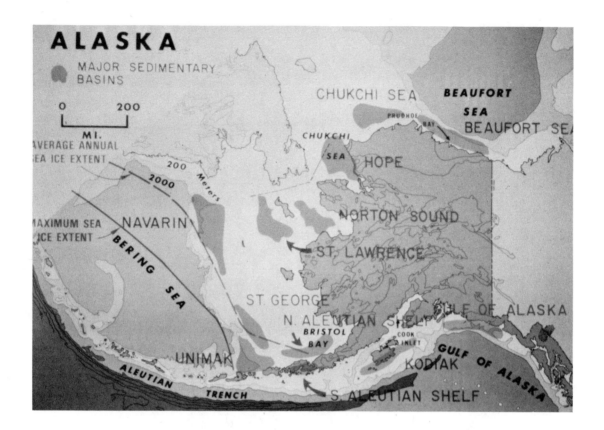

Figure 2

Platforms in Existing Lease Areas

Cook Inlet

Oil field development in the Upper Cook Inlet dates back to the early 1960's. A total of 14 platforms were built there between 1964 and 1968, and the first of these have now a 15-year history of successful operation in sea ice. Some unique designs like the monopod structure in Figure 3,[1] were developed to cope with a formidable combination of environmental factors, including 30 foot tides, 8 knot currents, earthquakes, and sea ice.

Designing against the fast-moving ice floes was clearly the most difficult task in view of the fact that there was little prior experience on which to base a safe design. Extensive laboratory and field testing was conducted to determine maximum ice loads that would have to be resisted. This research program included in situ measurements of ice loads on instrumented test piles (Figure 4)

Figure 3

Figure 4

attached to a temporary structure.[2] It provided the data needed for safe design of the production platforms that were built in the following years.

The four-legged Cook Inlet platforms (Figure 5) were designed for an ice thickness of 3.5 feet and effective ice pressure of 300 psi acting on all four legs simultaneously.[3] This translates into a total ice load of nearly ten million pounds. Several of the Cook Inlet platforms have been instrumented to measure actual ice loads in subsequent winters.[4,5] The measurements indicated that the effective ice pressure experienced by these structures was less than 125 psi, i.e., less than half of the design value of 300 psi.[5] This indicates that there is an adequate margin of safety in the design.

All of the Cook Inlet structures are characterized by large caisson-like legs and an elaborate piling system. To minimize exposure to the sea ice, all wells are drilled through the legs, eliminating external well conductors, and there are no braces between the legs in the vicinity of the water level. The structure in Figure 5 is held in place with 32 double-walled grouted piles, eight in each leg, to resist the large shear loads and overturning moments resulting from the crushing force of the sea ice. The total weight of the piling system (about 4,300 kip) is significantly larger than that of the jacket (2,900 kip). The piles serve also as conductor guides for the production wells. The Cook Inlet platforms were installed in water depths up to 130 feet at high tide.

Pile-founded structures such as those developed for Cook Inlet may find application in other sub-Arctic sea ice areas such as Bristol Bay, where environmental conditions are similar to those in Cook Inlet.

Beaufort Sea

Let me turn now to the Beaufort Sea. The design approach just described for Cook Inlet served also as a model for the Arctic, even though the design criteria developed for Cook Inlet clearly do not apply in the Beaufort Sea: the Arctic ice is both thicker and stronger.

Field tests have been conducted near Tuktoyaktuk in Mackenzie Bay to measure directly the crushing strength of Arctic ice (Figure 6): large vertical cylinders were frozen into the ice in pairs, then forced apart by hydraulic jacks. The ice strength measured in up to five feet of ice ranged from 400 to 900 psi.[6] Significantly, the lower values were measured with the larger cylinders, five feet in diameter. This observed size effect was predicted on the basis of theoretical considerations and has been confirmed by subsequent tests with still larger ice crushing devices up to 12 feet wide.

Figure 5

Figure 6

These experiments formed the basis for the design of a monopod structure (Figure 7) for exploratory drilling which has received "Approval in Principle" by the Canadian government.[7] With a column diameter of 30 feet, this structure is designed for ice loads in excess of 20 million pounds, more than twice the design load of the typical four-legged structure in Cook Inlet. It was intended for use in water from 10 to 40 feet deep in Mackenzie Bay.

Figure 7

However, the monopod was never built. The reason is that man-made islands were technically and economically more attractive for exploration drilling in shallow arctic waters. Island technology was developed in the Canadian Beaufort Sea, where 17 islands have been constructed to date, most of them by Exxon's Canadian affiliate, Esso Resources Canada (ERC). The locations of these islands are shown on the map in Figure 8, including two, Unark and Pelly, built by Sun Oil Company. The first island, Immerk, was built in 10 feet of water in 1972; the latest, Issungnak, completed this fall, in 63 feet of water. Four islands were built at Adgo, where ERC discovered oil and gas. Most of these islands were constructed in the summer by dredging. Others were built in winter by trucking gravel over the ice and dumping the gravel through a hole excavated in the ice sheet.

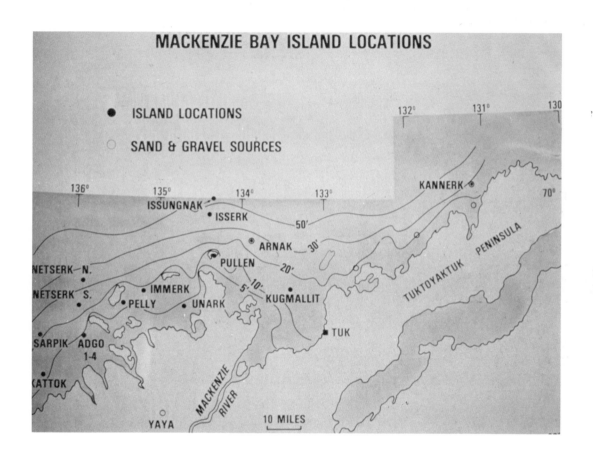

Figure 8

Man-made islands (Figure 9) offer the distinct advantage that drilling can be conducted in essentially the same manner as on land. They can be designed for year-round operations. Islands are gravity structures that resist lateral ice loads by their large weight. By adjusting the island size and freeboard, the sliding resistance on the sea floor, or on any given shear plane through the island fill, cane adjusted as necessary to assure a stable platform for the anticipated ice loading conditions. Thus, this type of structure can be easily adapted to site-specific design parameters. Also, temporary islands for exploration drilling can conveniently be enlarged and transformed into a permanent production platform if a discovery is made.

Figure 9

Gravel islands have been found to have minimal impact on the environment, both during construction and after the islands have been completed.[8] Once abandoned, they disappear gradually due to natural erosion without harmful environmental effects (Figure 10).

Figure 10

Design considerations and construction techniques for gravel islands in Arctic environments have been well-publicized in the technical literature. A number of papers have been published[9-14] dealing with specific island designs and operating experience, in both summer and winter. Extensive research has been conducted, both in the laboratory and in the field, to define sea ice conditions and ice properties in the Alaskan Beaufort Sea, and to develop safe design criteria for various ice loading conditions.[15-20] Exxon USA has recently conducted a public seminar on this subject, wherein Exxon USA discussed design procedures for gravel islands in the Alaskan Beaufort Sea. This technology is directly applicable to the majority of the lease area to be offered in the federal/state lease sale in the Alaskan Beaufort Sea, scheduled to be held next week. Several gravel islands have already been built on existing state leases, in water depths up to 11 feet (Figure 11).

One unique aspect of gravel island operations has been the development of active island defense and monitoring systems. The temporary nature of exploration islands places a high penalty on over-designing, in terms of both cost and gravel use. Also, because of the relatively short-term operation, the extra cost of active

Figure 11

defense methods, monitoring and perhaps shutdowns in case of unusual
ice conditions, can be tolerated without excessive economic
consequences. Active defense measures include weakening, by
slotting, of certain ice types and features such as thick floes of
multi-year ice or rafted ice; real-time monitoring of ice movements
and ice pressures; and emergency well shut-in procedures. An example
of instrumentation for measuring in-situ ice pressure is shown in
Figure 12. This photograph was taken at Exxon's Duck Island location
in the Alaskan Beaufort Sea.

In the shallow Arctic environment, the sea ice cover becomes
stabilized during early winter and undergoes only slow deformations
during the rest of the winter. Thus, in late winter, when ice loads
are potentially the most severe, the rate of pressure changes is very
slow, with time scales in the order of hours or days. By monitoring
these slow ice deformations and pressure changes, warnings can be
provided with sufficient time for response action and contingency
measures before the ice forces reach the island's design strength
level.

As I mentioned, the industry has, so far, built only temporary
islands that were used for exploration drilling. For permanent
production islands, we anticipate that a passive design against
lateral ice loads will be required. Active defense measures, other
than monitoring of ice conditions, will not be economically or
operationally attractive for long-term, year-round production
operations. Thus, the design of production islands will have to

Figure 12

-----

accommodate more extreme ice events than is the case for temporary
islands. This can be accomplished by increasing the size and
freeboard of the island.

Next, I want to comment on a different type of drilling platform,
artificial ice islands, that may be a viable alternative to gravel
islands in some shallow-water locations. During the winter of 1976-
1977, Union Oil drilled an exploration well from a grounded ice
island (Figure 13) built in 9 feet of water in Harrison Bay, about 50
miles west of Prudhoe.[21] Construction of the Union ice island
began in early November by repeated flooding of the ice surface with
seawater. In this fashion, the ice was gradually thickened and
eventually grounded to form a platform for the drilling operation.
Island construction was completed by January 20 when the platform had
a freeboard of about three feet. Then the rig was moved on and the
well drilled in time to vacate the island by mid-April. The ice
island melted and disappeared in early July, shortly after breakup.

Figure 13

Because of their low weight, low-freeboard ice islands lack the stability of gravel islands. Therefore, they need to be protected against lateral movement of the surrounding ice. Union Oil protected its ice island by maintaining an 11 foot-wide moat (Figure 14) around most of the island. This separated the island from the surrounding ice sheet so that any motion of the ice sheet would not exert forces on the island which might push it off its location. The moat was opened by ditching machines, cutting the ice into blocks, and a crane which lifted them out to the water.

Exxon, in cooperation with three other companies (Mobil, Phillips and Sohio), has conducted a prototype experiment to extend ice island technology. During the last winter, a large experimental ice island (Figure 15) was constructed in 10 feet of water north of Prudhoe Bay. A modified center-pivot irrigation system (Figure 16) was used during part of the experiment. The diameter of the island was about 1,200 feet, and it achieved more than 20 feet freeboard at the center. The experiment had the following objectives:

1.   To test improved techniques for rapid, streamlined ice island construction.

2.   To demonstrate the concept of a large ice island with sufficient mass to resist lateral ice forces during spring breakup.

Figure 14

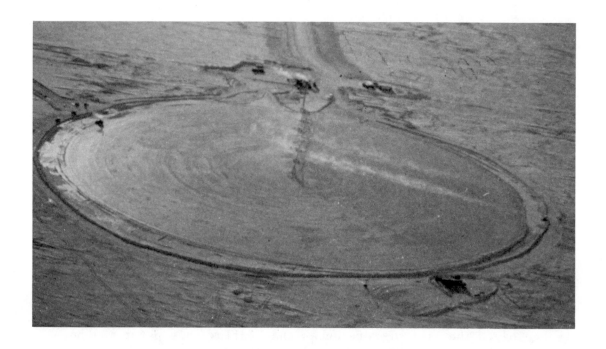

Figure 15

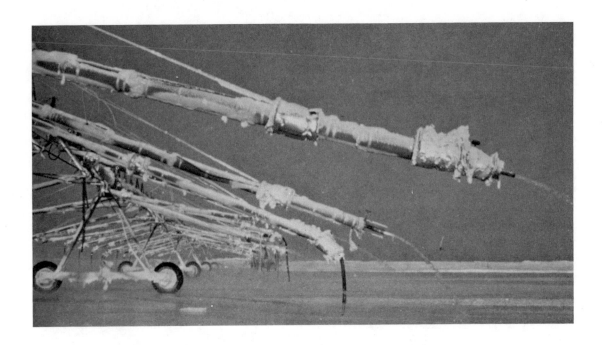

Figure 16

---

3. To observe the rate of deterioration of a
large ice island during the summer open-
water season.

The experiment accomplished all three objectives. We produced a
massive island that remained on location through breakup (Figure 17).
Deterioration during the summer was found to be significant, and the
island disappeared before the end of summer. This means that some
sort of melt protection is required if we want to plan for an ice
island to be available for a second winter drilling season. The
feasibility of preserving an ice island through the open-water summer
season has not been demonstrated. In any case, the need for pro-
tective measures would add to the cost and reduce the economic
attractiveness of two-year ice island concepts.

This concludes my review of industry experience with ice-
resistant platforms in existing lease areas. In this limited time, I
could only touch on the technology that has been developed during the
last 15 years to deal successfully with ice conditions in both sub-
Arctic (Cook Inlet) and Arctic (Beaufort Sea) sea ice conditions.
However, there have been numerous publications dealing with this
technology. The list of references that follow this paper may be
useful to those who want more detailed information.

Figure 17

Platforms for Future Lease Areas

Let me turn now to future lease areas in ice-covered waters of the Alaskan OCS. Some new platform types will be needed to accommodate the various combinations of environmental conditions--sea ice, water depth, earthquakes, and soil conditions--that are present in these areas. In many cases, structures can be developed by adaptation or extensions of existing technology, such as the pile-founded structures in Cook Inlet or concrete gravity structures in the North Sea. In general, the hostile environment and remoteness of the sea ice areas favor the use of gravity structures that can be preassembled at a construction site in temperate waters, towed to location, and installed quickly with all or most of the production facilities already in place.

Beaufort Sea

I will begin with the Beaufort Sea. The upcoming lease sale will extend to about 60 feet of water in a relatively small area stretching from the vicinity of Prudhoe Bay in the west to the Canning River in the east. As I mentioned before, gravel islands appear to be the most generally applicable drilling platform for this lease area. Exxon's Canadian affiliate has constructed islands in water depths up to 63 feet in Mackenzie Bay. Off the North Slope, the shorter open-water season for summer dredging tends to limit the depth range of man-made islands for economic reasons, but we consider gravel islands practical today out to a water depth of at least 40 feet. Many of the leases that are located in water deeper than

40 feet could be reached by directional drilling from locations along the 40 feet depth contour. This is true for exploration drilling as well as for development drilling. Thus, man-made islands are also the most likely concept to be used for production platforms if a discovery is made in this lease area.

As the water depth increases, the construction of gravel islands becomes increasingly difficult and costly, and other types of drilling platforms may become more attractive. Exxon and other oil companies are conducting research and development work on alternative drilling methods that would be applicable in deeper Arctic waters. Chief among these are mobile conical gravity structures which are designed to break the ice in bending rather than crushing, at significantly reduced loads. Both concrete and steel structures have been proposed.

Esso Resources Canada developed a preliminary design for a conical concrete structure (Figure 18) for use in Mackenzie Bay in

---

Figure 18

water depths up to 135 feet.[22]   The structure consists of a large-diameter circular hull, a cone section with a 45 degree cone angle and 40 ft. diameter at the top, and a multi-story deck section.   The hull is designed for impact by deep-keeled ice features and serves two main functions: (1) to provide resistance against sliding and overturning when the structure rests on bottom; and (2) to provide buoyancy when deballasted so that the structure can be towed while floating on its hull in a stable configuration and with minimum draft.   This particular structure was designed for operating in the depth range from 70-135 feet.   It has a hull diameter of 450 feet.   A smaller cone structure for 35-70 feet water depth, with a hull diameter of 330 feet, was also designed.

As I mentioned, these gravity structures are intended for exploration drilling.   Their depth range capability is limited by the requirement that the cone must be in a certain position relative to the water level, to miminize ice loads.   This is a severe constraint for an exploration concept, limiting its applicability to a narrow depth range in the particular ice environment for which it is designed.

The monocone concept (Figure 19), also developed by Esso Resources Canada, would mitigate this limitation somewhat by the use of a movable conical collar that can be adjusted vertically on a cylindrical shift to the optimum position for ice breaking.[23] Advantages over structures with a fixed cone designed for the same depth range are (1) reduced weight; and (2) reduced cone surface area where heating or special coatings may be required to prevent ice from freezing to the structure.   Monocone designs have been developed for three water depth ranges: 35 to 70 feet, 70 to 135 feet, and 135 to 200 feet.

The cone and monocone designs developed by Esso Resources Canada are designed to resist large multi-year pressure ridges up to 45 feet thick.   This design criterion was judged to be adequate for applications in the Mackenzie Bay area, where the frequency of multiyear ice invasions with heavy consolidated pressure ridges is relatively low.   However, this design criterion may not be sufficient for the North Slope, where grounding multiyear ridges occur in water depths out to at least 60 feet.   In order to reduce potential ice loads developed from interactions with such large multiyear pressure ridges, cone angles of less than 45 degrees may be considered.

Figure 19

Extensive model studies have been conducted to determine the
ridge loads that are developed on cones with different slope angles.
For example, Exxon conducted model tests last winter, with joint-
industry funding, in ERC's outdoor test basin in Calgary, Alberta,
The model cone (Figure 20) had a 30 degree slope and was approximately one-tenth scale. The model tests simulated interactions with
partially consolidated ice rubble and with fully consolidated
pressure ridges (Figure 21). Additional model tests are planned for
this coming winter. The results of these tests will form the basis
for improved cone structure designs for the Alaskan Beaufort Sea.
Other model tests have been conducted to measure wave loads and to
observe the floating stability of cone structures during tow and
installation.[24] The upcoming Beaufort Sea lease sale, if held as
planned, will provide strong incentives to further develop and field
test such advanced drilling concepts for the Arctic environment. A
feasibility study for a conical test structure, for field testing in
the Beaufort Sea, has recently been initiated with joint industry

Figure 20

Figure 21

funding. If gravity structure concepts are developed for exploration drilling in the 40 to 60 foot depth range in the proposed federal/state Beaufort Sea lease area, it is anticipated that these concepts can be extended to deeper water, perhaps 100 feet, for both exploration and production, by the end of the next decade.

Chukchi and Bering Sea

As I mentioned before, the Beaufort Sea and perhaps the northern Chukchi Sea are the only areas among those scheduled for leasing in the five year lease plan where new structure concepts may be needed for exploration drilling in water depths beyond the economic limits of artificial gravel islands. Conventional open-water drilling will be feasible in proposed lease areas in the southern Chukchi and Bering Sea.

Ice-strengthened drilling vessels, in combination with various techniques for breaking and diverting ice floes, may be employed in parts of the Chukchi Sea in order to extend the summer drilling season beyond the open-water period. This approach is currently being used in deeper portions (more than 100 foot water depth) of the Mackenzie Bay area in the Canadian Beaufort Sea.[25]

However, gravity structure concepts developed for exploration drilling in the Beaufort Sea can probably be adapted for production platforms in the less severe ice conditions of the southern Chukchi Sea and northern Bering Sea. Thus, we anticipate that cone gravity structures would be a prime candidate for production platforms in the Hope and Norton basins and perhaps the St. Lawrence basin, in water depths up to perhaps 200 feet. Man-made islands would, of course, be applicable in the shallow-water portions of these basins.

The required design strength levels in these areas are expected to be similar to or less than those in Mackenzie Bay, for which preliminary designs have already been developed for water depths up to 200 feet. Several industrial research projects have recently been undertaken to study sea ice conditions in the Bering Sea in detail.[26,27] Detailed platform geometry and design would, of course, be tailored to the specific requirements (regarding the number of wells and production rates) and local environmental conditions, particularly with respect to foundation soil properties, at the site of a discovery.

In the southern part of the Bering Sea, as ice conditions become less severe, the design loads will become small enough so that adaptations of more conventional platform designs will become practical. Both pile-founded and gravity-type structures can be considered for these areas. The open-water summer season is of sufficient length to allow extensive pile-driving operations.

Therefore, pile-founded caisson structures, similar to those used
in Cook Inlet, can probably be designed for the relatively shallow
Bristol Bay area. Environmental conditions there are similar to
those of Cook Inlet.

Farther to the west, in the St. George and Navarin basins, the
deeper water and the remoteness of the area will tend to favor
gravity-type structures, which can be installed with a minimum of
on-site construction and with most or all of the deck facilities
already in place. Exxon has conducted several studies to assess the
applicability in the Bering Sea of tower-shaped gravity structures,
similar to those that are currently in use in the North Sea.

The water depth of interest ranges from 300 feet to about 650
feet. Earthquakes are much less severe than to the south of the
Aleutian chain, but they are nevertheless an important design
consideration.[28] The expected ice conditions vary from
occasional scattered drift ice in the southern portions of the
St. George area to 5-foot thick floes of annual ice and
unconsolidated ridges 60 feet in the northern Navarin basin. Soil
conditions will also vary widely in an area of this size. No single
platform configuration will, therefore, be suitable for all
combinations of sea ice conditions and soil properties to be
encountered in this large region. I will briefly discuss three
different platform geometries that we judge to have potential
applicability and that together appear to span the principal range
of design conditions expected in the St. George and Navarin basins.

The first platform type is a rather conventional concrete
gravity structure similar to the three-legged Condeep structures
(Figure 22) in the North Sea. These structures have been designed
for the severe North Sea wave climate which produces large lateral
loads. We found that structures of this type could also resist
significant ice loads with only minor design modifications. Their
principal limitation is that firm foundation soils would be required
to resist (1) the large overturning moment developed by sheet ice
crushing against the three legs, and (2) the large base shear forces
induced by earthquakes. Given adequate soil conditions such as firm
sand or stiff clay, we anticipate that this type of structure can
be designed for water depths from 300 to 500 feet in the St. George
area. The environmental conditions assumed are listed on Figure 23.
They include 3 foot sheet ice, 30 foot unconsolidated pressure
ridges, an 86 foot design wave, and .2g earthquake acceleration. Ice
forces were calculated by assuming 450 psi effective ice pressure
for sheet ice crushing and 5 psi cohesive strength for ridge shear-
ing,[29] both occurring at the same time on all three platform legs.
We believe these are conservative assumptions for the St. George
area. We found that, with certain modifications, Condeep-type
structures could be designed to these conditions for clay soils with
2 ksf or more cohesive strength or sands with a friction angle of
about 35 degrees or more.

Figure 22

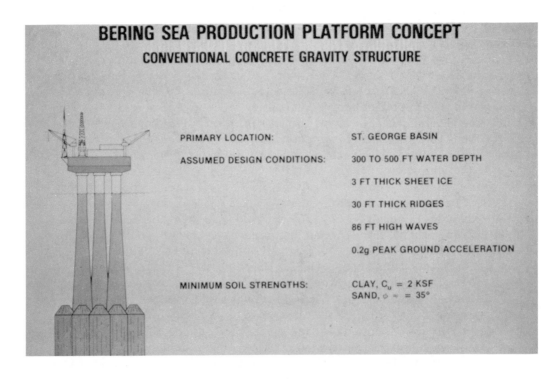

Figure 23

Several modifications of this basic platform design have been considered in order to extend the applicability to softer soils or more severe ice conditions. Two ways of reducing potential ice loads and the resulting foundation loads are: reducing the number of platform legs, and attaching an ice-breaking cone near the waterline. Our studies along these lines have resulted in two platform types that are judged to be applicable in the main portions of the St. George and Navarin basins.

The first of these is the monotower concept shown in Figure 24. It would be applicable in most of the St. George basin area, with the environmental conditions mentioned previously, but with softer soils, 1 to 1.5 ksf cohesive strength, depending on water depth, or sand with 32 degree friction angle. Platforms of this type could also be designed for shallow portions of the Navarin basin, where ice conditions are somewhat more severe than in the St. George area. As many as 30 production wells could be drilled through the single support tower of this structure.

The second concept, a monotower with ice-breaking cone (Figure 25), may be used in deeper portions of the Navarin basin. The cone section near the waterline is designed to cause the advancing ice sheet to break in bending rather than crushing at significantly reduced load.[30] This results in lower overturning moment, a critical design parameter in deep water. The environmental parameters assumed for this design are listed on Figure 25: 300-650 foot water depth, 4 foot sheet ice, 60 foot pressure ridges, 90 foot waves, and .1g earthquake acceleration. This is only half the value assumed in the St. George area. The ice loads were calculated for bending failure of the 4 feet sheet ice, assuming 100 psi flexural strength, and simultaneous shear failure of the pressure ridge, assuming again 5 psi cohesive strength. The minimum required soil cohesive strengths were found to be again in the range from 1 to 1.5 ksf, depending on water depth, or sand with friction angles of 32 degrees or better. A perspective view of this monotower structure is shown on Figure 26.

I need to stress again that the platform types I have shown are the result of parameter studies that attempted to span the environmental and soil conditions anticipated in the St. George and Navarin basins which comprise the deeper portions of the Bering Sea shelf. They are not necessarily the only types of bottom-founded platforms to be considered in these areas, and the optimum designs for specific locations may look different from the ones I have shown. The important point is that our studies have identified feasible and practical production platform concepts for each of the various sedimentary basins identified on the Bering Sea shelf.

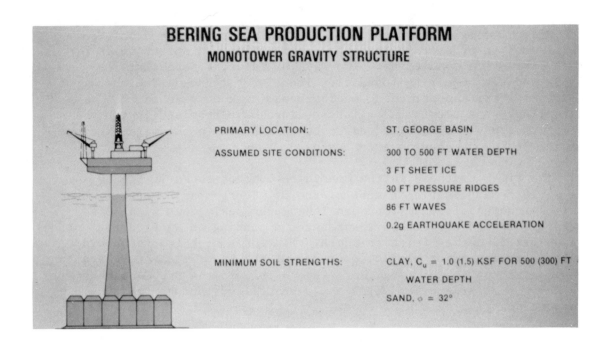

BERING SEA PRODUCTION PLATFORM
MONOTOWER GRAVITY STRUCTURE

PRIMARY LOCATION: ST. GEORGE BASIN

ASSUMED SITE CONDITIONS: 300 TO 500 FT WATER DEPTH

3 FT SHEET ICE

30 FT PRESSURE RIDGES

86 FT WAVES

0.2g EARTHQUAKE ACCELERATION

MINIMUM SOIL STRENGTHS: CLAY, C_u = 1.0 (1.5) KSF FOR 500 (300) FT

WATER DEPTH

SAND, ϕ = 32°

Figure 24

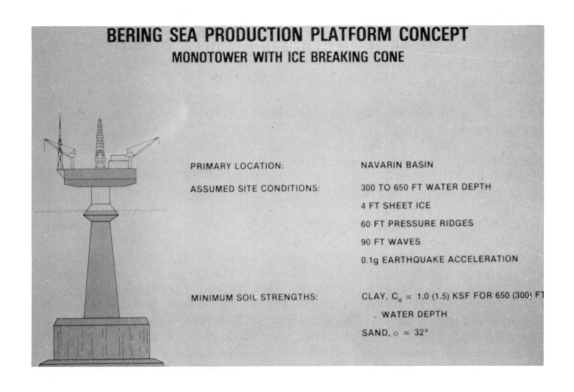

BERING SEA PRODUCTION PLATFORM CONCEPT
MONOTOWER WITH ICE BREAKING CONE

PRIMARY LOCATION: NAVARIN BASIN

ASSUMED SITE CONDITIONS: 300 TO 650 FT WATER DEPTH

4 FT SHEET ICE

60 FT PRESSURE RIDGES

90 FT WAVES

0.1g EARTHQUAKE ACCELERATION

MINIMUM SOIL STRENGTHS: CLAY, C_u = 1.0 (1.5) KSF FOR 650 (300) FT

WATER DEPTH

SAND, ϕ = 32°

Figure 25

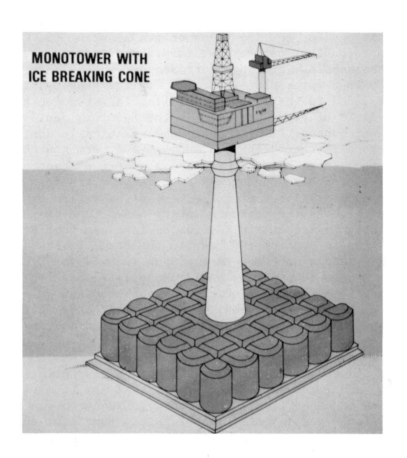

Figure 26

Conclusion

In conclusion, I would like to summarize our assessment of Arctic platform technology for sea ice areas on the Alaskan OCS. Existing gravel island technology can be used for year-round exploration drilling in the shallow-water Beaufort Sea out to at least 40 feet and perhaps 60 feet water depth. Mobile conical gravity structures are being developed for exploration drilling in water depths beyond the economic limit of gravel islands in the Beaufort Sea. Conventional offshore drilling techniques are suitable during the open-water season in the Bering Sea and the southern Chukchi Sea. If discoveries are made in any of these areas, ice-resistant platforms will be required for development and production operations. Man-made production islands will be applicable in shallow waters of the Beaufort, Chukchi, and the northern Bering Sea. Cone-shaped production platforms are likely candidates in these same areas in deeper water, out to perhaps 100 feet in the Beaufort Sea, 200 feet in the Chukchi and the northern Bering Sea. Cook Inlet-type structures with pile foundations could probably be used in the Bristol Bay area. And finally, various adaptations of gravity tower concepts, proven in the North Sea, would be applicable on the deeper portions of the Bering Sea shelf, in the St. George and Navarin basins.

While the industry will continue to improve and optimize these designs, and to develop new concepts that may be more attractive than the ones I have shown today, I want to emphasize that we have, for several years, investigated structural designs for sea ice areas, and have identified feasible and practical concepts for each of the various sedimentary basins in the sea ice areas of the Alaskan OCS.

The industry has learned to cope with sea ice in Cook Inlet and in those portions of the Beaufort Sea that have already been leased. The engineering tools exist to design structures for other sea ice areas as well. Suitable platform types have been identified. We can begin to design platforms for site-specific environmental conditions, and for specific operational requirements, as soon as a discovery is made.

Dr. Jahns is Senior Research Advisor of Exxon Production Research Company, and served on the Marine Board's Panel on Polar Ocean Engineering. Educated at Technical University, Clausthal, Germany, he holds degrees in Geology (Diploma), Mining Engineering (Diploma), Petroleum Engineering (Dr. Ing). He is a member of the National Research Council's Polar Research Board and Permafrost Committee of the Polar Research Board, as well as a member of the Advisory Board of the Geophysical Institute of University of Alaska.

REFERENCES

1. Cloid, M. P., "Monopod," Civil Engineering - ASCE 38, pp. 55-57, March 1968.

2. Peyton, H. R., "Sea Ice Forces," Ice Pressures Against Structures, (Proc. of a Conf. at Laval Univ., Quebec, Nov., 1966) NRC Tech Memorandum 92, pp. 117-123, March 1968.

3. Visser, R. C. "Platform Design and Construction in Cook Inlet Alaska," Journal of Petroleum Technology, pp. 411-420, April 1969.

4. Geminder, R., "Ice Force Measurement," Proceedings of OECON - Offshore Exploration Conference, pp. 309-336, New Orleans, La., February 1968.

5. Blenkarn, K. A., "Measurement and Analysis of Ice Forces on Cook Inlet Structures," OTC 1261, Preprints, Offshore Technology Conference, pp. II 365-378, Houston, Texas, 1970.

6. Croasdale, K. R., "Crushing Strength of Arctic Ice," Proceedings of a Symposium on Beaufort Sea Coast and Shelf Research, Arlington, Virginia, Arctic Institute of North America, pp. 337-399, 1974.

7. Jazrawi, W., and Davies, J. F., "A Monopod Drilling System for the Canadian Beaufort Sea," S.N.A.M.E., Ice Tech '75, Montreal, 1975.

8. ___, "Artificial Islands in the Beaufort Sea--A Review of Potential Environmental Impacts," Fisheries and Marine Service, Environmental Secretariat, D. C. Wright, Coordinator, Canada, September 1977.

9. deJong, J.J.A., and Bruce, J. C., "Design and Construction of a Caisson Retained Island Drilling Platform for the Beaufort Sea," Preprints Offshore Technology Conference, Houston, Texas, pp. 2111-2120, May 1979.

10. Croasdale, K. R., and Marcellus, R. W., "Ice and Wave Action on Artificial Islands in the Beaufort Sea," Canadian Journal of Civil Engineering, Vol. 5, No. 1, pp. 98-133, 1978.

11. Garratt, D. H., and Kry, P. R., "Construction of Artificial Islands as Beaufort Sea Drilling Platforms," Journal of Canadian Petroleum Technology, pp. 73-79, April-June 1978.

12. Riley, J. G., "The Construction of Artificial Islands in the Beaufort Sea," Journal of Petroleum Industry, pp. 365-371, April 1976.

13. deJong, J.A.A., Stigter, C., and Steyn, B. "Design and Building of Temporary Artificial Islands in the Beaufort Sea," Proceedings, Third POAC Conference, pp. 753-789, Fairbanks, Alaska, 1975.

14. Brown, A. D. and Barrie, K.W., "Artificial Island Construction in the Shallow Beaufort Sea," Proceedings, Third POAC Conference, pp. 705-717, Fairbanks, Alaska.

15. Wang, Y. S., "Crystallographic Studies and Strength Tests of Field Ice in the Alaskan Beaufort Sea," Proceedings, Fifth POAC Conference, pp. 651-655, Trondheim, Norway, 1979.

16. Templeton, J. S. III, "Measurement of Sea Ice Pressure," Proceedings, Fifth POAC Conference, pp. 73-87, Trondheim, Norway, 1979.

17. Prodanovic, A., "Field Ice Strain Measurements," Proceedings, IAHR Symposium on Ice Problems, pp. 151-164, Lulea, Sweden, 1978.

18. Wang, Y. S., "Buckling Analysis of A Semi-Infinite Ice Sheet Moving Against Cylindrical Structures, Proceedings, IAHR Symposium on Ice Problems, pp. 117-133, Lulea, Sweden, 1978.

19. Ralston, T. D., "An Analysis of Ice Sheet Indentation," Proceedings, IAHR Symposium on Ice Problems, pp. 13-31, Lulea, Sweden, 1978.

20. Kry, P. R., "A Statistical Prediction of Effective Ice Crushing Stresses on Wide Structures," Proceedings, IAHR Symposium on Ice Problems, pp. 33-47, Lulea, Sweden, 1978.

21. ___, "Union's Beaufort Sea Ice Island Success," Oil and Gas Journal, pp. 42-43, July 11, 1977.

22. Croasdale, K. R., "Ice Forces on Fixed Rigid Structures," Report prepared for the Working Group on Ice Interaction on Hydraulic Structures, Committee on Ice Problems, International Association for Hydraulic Research, April 1978.

23. Jazrawi, W. and Khanna, J., "Monocone - A Mobile Gravity Platform for the Arctic Offshore," Proceedings, Fourth POAC Conference, St. John's, pp. 170-184, Newfoundland, 1977.

24. Ibid.

25. ___, "Dome Winds Up Beaufort Sea Drilling," Oilweek, p. 9, November 5, 1979.

26. Deily, F. H., "Aerial Reconnaissance and Subsea Profiling of Sea Ice in the Bering Sea," Proceedings, Fifth POAC Conference, pp. 207-219, Trondheim, Norway, 1979.

27. Barnes, J. C. and Bowley, C. J. "A Five-Year Sea Ice Climatology of the Bering Sea Derived from Satellite Observations," Proceedings, Fifth POAC Conference, pp. 191-205, Trondheim, Norway, 1979.

28. ___, "Offshore Alaska Seismic Exposure Study," prepared by Woodward-Clyde Consultants for the Alaska Subarctic Offshore Committee, San Francisco, California, March 1978.

29. Prodanovic, A., "Model Tests of Ice Rubble Strength," Proceedings, Fifth POAC Conference, pp. 89-105, Trondheim, Norway, 1979.

30. Ralston, T. D., "Ice Force Design Considerations for Conical Offshore Structures," Proceedings, Fourth POAC Conference, pp. 741-752, St. John's Newfoundland, 1977.

OFFSHORE DISCHARGE OF DRILL MUDS AND CUTTINGS

James P. Ray

Since the turn of the century, rotary drilling has been the predominant drilling technique used to drill oil wells. A drill bit (of which there are innumerable types) is pressed against the bottom of the bore hole, and under the considerable weight and rotary motion imparted to it by the drill pipe, chips, crushes, or grinds the formation rock. Integral to the modern drilling techniques are the use of various types of fluids to aid in the removal of cuttings as well as perform a broad variety of downhole functions to optimize the drilling process. Table 1 gives approximate percentages of the different types of drilling fluids currently being used. Water base drilling fluids are by far the most commonly used fluids in offshore drilling. In these muds, water is the continuous phase.

Drilling muds serve a number of essential functions (Figure 1). Foremost is the maintenance of well control by providing sufficient hydrostatic head to counterbalance formation pressures encountered down hole. Increasing pressure gradients are counterbalanced by increasing mud weight (normal mud weights range from 9 to 20 pounds per gallon). Also very important is the removal and transport of drill cuttings to the surface. Additional functions include: cool and lubricate the bit, control water loss, bore hole stabilization; and to buoy the drill pipe and casing.

Table 1

Air and Gas	2.0%
Oil	1.3%
Oil Base Emulsion	2.6%
Oil Base	5.3%
Clear Water	9.1%
Water Base Muds	79.7%

DRILLING FLUID CATEGORIES AND PERCENTAGE USAGE

The composition of drilling fluids varies widely and is determined by the downhole conditions encountered and the specific requirements of the company drilling the hole. In the early stages of the hole, sea water and clay are the primary constituents. At increasing depth, clays and/or polymers are added to control viscosity and filter loss properties.

Figure 1/Mud Flow and General Mud Functions

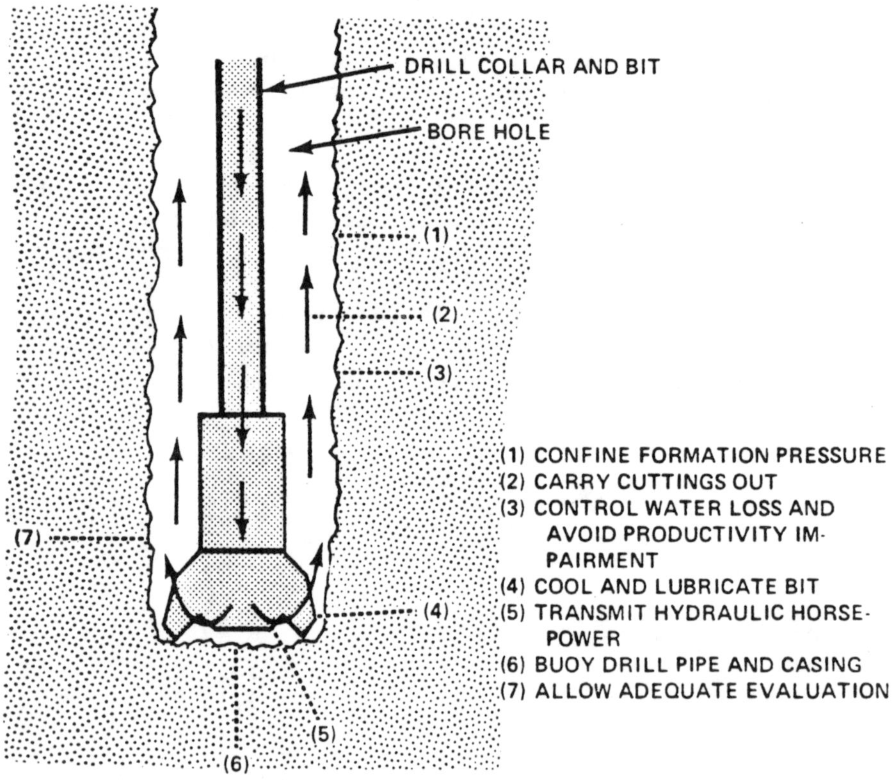

DRILL COLLAR AND BIT

BORE HOLE

(1)

(2)

(3)

(7)

(4)

(5)

(6)

(1) CONFINE FORMATION PRESSURE
(2) CARRY CUTTINGS OUT
(3) CONTROL WATER LOSS AND
     AVOID PRODUCTIVITY IM-
     PAIRMENT
(4) COOL AND LUBRICATE BIT
(5) TRANSMIT HYDRAULIC HORSE-
     POWER
(6) BUOY DRILL PIPE AND CASING
(7) ALLOW ADEQUATE EVALUATION

When weight is needed in the system, barite is the material
usually added. A variety of chemicals are added in minor amounts to
control various mud properties such as viscosity, filter loss, and
cation (e.g., Ca, Na, and Mg) contamination. In very general terms,
a typical freshwater mud is 1-5 percent active clay, 0-40 percent
inert solids, and the remainder water (volume by %). A detailed
discussion of the various "types" of muds and their constituents is
beyond the scope of this paper. Although there are more than 500
trade name mud products, chemically, they are represented by about 55
different compounds, of which only around a dozen are used on any one
well.

Drill Mud Systems

The drill mud system is an integral part of any drilling rig. A
typical "mud system" can be seen in Figure 2. A series of mud tanks
are contained in the area adjacent to the drill floor. These tanks
sometimes hold mud of several weights for use at increasing depths.
Typical rig holding capacity can range from approximately 1,000
barrels (bbl) to 2,500 bbls (1 bbl = 42 gallons). The mud then goes
to a slugging tank where additional materials can be added as needed.
From here, the mud is pumped at high volume (200-600 gallons per
minute) and pressures (1,200-3,500 psi) to the mudhose which leads
to the "kelly" (drive section of drill pipe) on the drill floor.
The mud then passes down through the drill pipe and out through
nozzles in the drill bit where the mud hydraulically assists in
the removal of cuttings.

The mud then passes up through the annulus (area between the
drill pipe and borehole or casing) carrying cuttings with it to the
mud return line. Mud and cuttings then flow onto a series of shale
shaker screens of varying mesh size. Cuttings are removed and fall
into the cuttings discharge hopper where they are washed down the
discharge pipe. The mud and small drill solids fall through the
shaker screen and are recirculated to the mud systems. Gas is
removed by a degasser. Desanders, desilters, mud cleaners, and
centrifuges remove specific size ranges of solids from the drill
fluid. The processed mud goes to the mud tanks and is ready to be
circulated downhole.

Drilling Discharges

One of the least understood phases of OCS operations to those
outside the industry is the discharge of drill muds and cuttings.
Many people have the impression that there is a steady rain of
drilling mud and cuttings falling from an offshore platform. To the
contrary, the discharge is quite intermittent, and the discharge
rates (volume and weight) are quite variable.

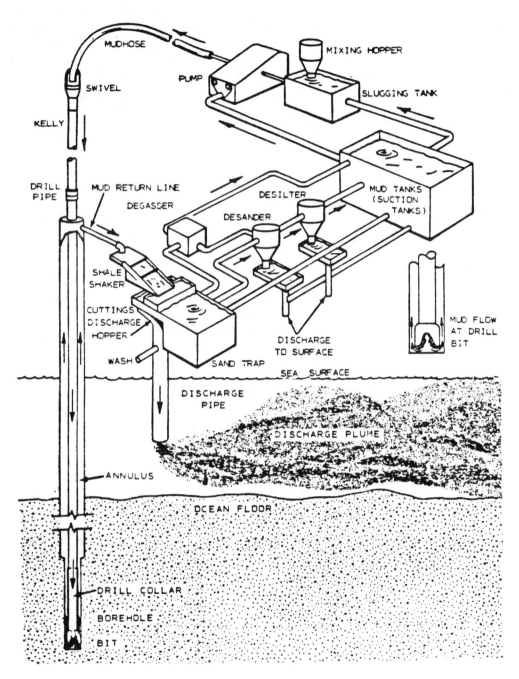

Figure 2

DRILLING FLUID CIRCULATION PATH

The typical well begins with the drilling or jetting of a surface hole (usually 30" diameter) to a depth of 100-350 feet. The materials resulting from this first several hundred feet are lost to the sea floor and not returned to the drilling rig. After surface casing is set and returns to the rig are established, progressive sections of deeper hole are drilled and smaller and smaller bits are used. Therefore, the actual volume of cuttings discharged steadily decreases with increasing well depth. For example, the well records from a Southern California exploratory well (Figure 3) show the relationship between the depth drilled and cubic feet of cuttings produced. The majority of cuttings discharged occurs in the early sections of the hole when large diameters are being drilled. As the hole goes deeper, the cubic feet of cuttings per day decreases.

During the drilling process, a number of different tasks can interrupt the actual drilling (and discharge) process. These include well testing, logging, cementing, well surveys, and drill bit replacement. Under average conditions, most drill bits have a life span of approximately 20-100 hours. When drilling pressures and penetration rates indicate that bit replacement is necessary, the drilling must be stopped, and the entire drill string pulled from the hole (usually dismantled in 90 foot "stands"). At deeper depths, this "tripping" procedure of removal, replacement of bit, and reentry can take twelve hours or more. In Figure 4 the intermittent nature of the drilling and the decreasing volume of cuttings discharged can be seen. During the three months required to drill this exploratory well, no drilling occurred one-third of the time.

Drilling fluids are discharged in two primary ways. First and most common is the loss of drill fluids associated with cuttings. As discussed in the previous Drill Mud Systems section, drill mud is returned to the surface carrying cuttings. The cuttings are removed by the shale shaker screens, and the mud is recirculated to the mud tanks and then back down hole. The volume of mud adhering to the discharged cuttings can vary considerably depending on the formations being drilled and the cutting size distribution. A general rule of thumb is 5 percent mud (by volume) associated with cuttings. As discussed for cutting discharge frequency, the releases are intermittent and decrease in volume with increasing hole depth.

Much less frequent is the bulk discharge of mud. This occurs several times during the life of a well and is usually done in order to create enough space in the mud tank(s) for the addition of extra diluent (make-up) water to change the mud properties. This usually is caused by the need to control solids content of the mud. Bulk discharges during normal drilling usually fall in the 100-300 barrel (bbl) range. A single bulk discharge of 300 bbls normally would last only 15-20 minutes. Bulk discharge frequency increases if dilution is

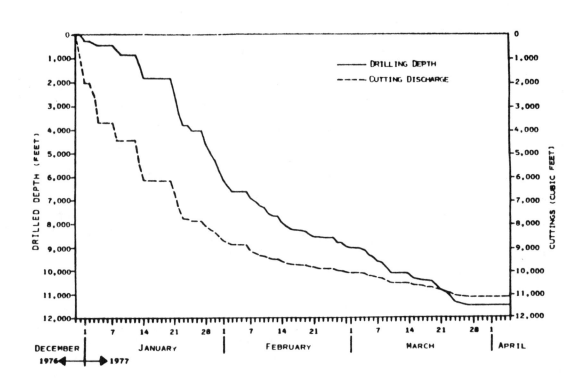

Figure 3

CUMULATIVE CUTTINGS DISCHARGE AND WELL DEPTH

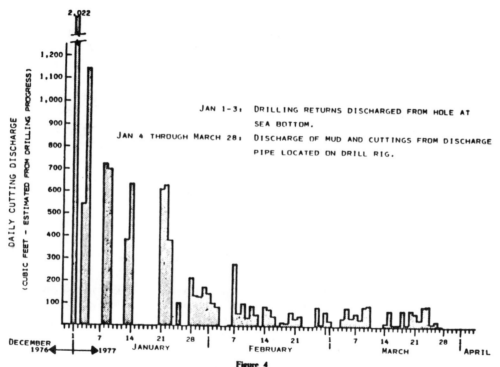

JAN 1-3: DRILLING RETURNS DISCHARGED FROM HOLE AT
          SEA BOTTOM.

JAN 4 THROUGH MARCH 28: DISCHARGE OF MUD AND CUTTINGS FROM DISCHARGE
          PIPE LOCATED ON DRILL RIG.

Figure 4
DAILY CUTTINGS DISCHARGE
January 1, to March 28, 1977

used as the primary solids control technique. Figure 5 shows the quantity and frequency of bulk discharges associated with the example exploratory well discussed in Figures 3 and 4. In exploratory drilling, due primarily to palaeontological contamination of the mud for future well evaluation, the muds are disposed of at the end of drilling. The bulk discharge usually ranges from 1,000-2,500 bbls. Depending on the discharge system, the entire dump can take from one to several hours. As depicted in Figure 5, two final bulk discharges were made, approximately 400 bbls on one day, and 1,200 bbls the next.

There is no standard discharge method for offshore drilling vessels. Jackup rigs, platforms, and semi-submersibles can be rigged in several ways. Discharge may be from the rig surface (50-100 feet above the water surface) and be a simple free fall to the water. Flexible hose may simply guide the mud and cuttings to the water surface. On other rigs, shunt pipes may be integral to the structure. These can vary in diameter (usually 12-18 inches) and can discharge anywhere from near the surface to within 5-10 meters of the bottom.

Recent studies on the actual behavior of mud and cuttings are giving us insight into future design of discharge systems for minimal environmental impact.

Discharge Regulations And Their Potential Cost

During the past few years there has been an increasing debate over the accepted discharge techniques for drill muds and cuttings. Although lacking the necessary scientific justification, regulatory agencies have pursued various avenues to control discharge. Some of the regulatory routes followed have included: setting discharge rates, not allowing bulk discharge, shunting materials near the bottom, and no discharge at all.

Industry is quite concerned with the trend of these regulations because a clear technical argument does not exist to justify the tremendous expense that could be added to the drilling of wells. Recent responses to proposed regulations have included some very rough approximations of the added costs. Potential costs accrue in many different ways, the first being in drill rig modifications for discharge or offloading. In addition, valuable deck space would have to be set aside for mud and cuttings storage between offloading periods. In relatively calm waters such as the Gulf of Mexico shipment of these materials can be handled by barge, which generally has a safe offloading condition of six feet. Thus the relative number of days per year unsuitable for the transfer operation would be small. As we move into more severe environments such as the Gulf

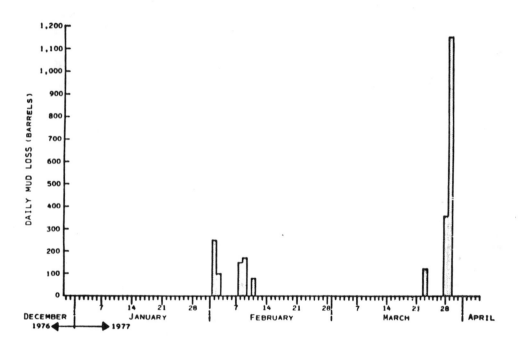

Figure 5

MUD RELEASES OTHER THAN THOSE ASSOCIATED'
WITH NORMAL CUTTINGS RELEASE

of Alaska or North Atlantic, we not only have a large number of "no offloading" days, but we have to use modified work boats for the hauling operation.

In addition to boat/barge costs, the daily costs for drilling rig time can be quite expensive, ranging from $35,000 to $125,000 per day. The large semi-submersibles and drillships used in exploratory frontier areas are in the $100,000 plus per day category. Due to the problems associated with obtaining ocean dumping permits from the Environmental Protection Agency (EPA), all materials would have to be transferred ashore, unloaded, and then transported to a suitable land disposal site.

The following are examples of some estimated costs. First, are potential costs for Gulf of Mexico operations at a drill site 100 miles offshore. The average well drilled in this area would be 10,000 feet and would take 35 days overall with 20 days actual drilling time and would generate 1,820 barrels of cuttings and 2,800 barrels of mud for disposal.

The cost of disposal, based on the lowest estimated cost would be:

Transport Cuttings ($1/bbl) (1,820 bbl)	$ 1,820
Transport Liquid Mud ($1/bbl) (2,800 bbl)	2,800
Dump Site Fee	
Cutting (20¢/bbl) (1,820 bbl)	364
Mud (20¢/bbl) (2,800 bbl)	560
Mud Cans for Storage (152 Cans) (5¢) (35 Days)	26,000
Offshore Boat ($140/hour) (35 days) (24 hours)	117,600
Crane Time ($25/hour) (40 hours)	1,000
Total	$150,744

Semi-Submersible

1. Disposal cost
   "Disposal of Mud and Cuttings" $150,744
2. Rig modification cost   30,000
3. Cost of rig downtime due to weather (assume 20 percent downtime per year, when seas are too rough to moor to rig; daily operating cost is $32,000 on stand-by basis. 7 Days x $32,000   224,000

Total $404,744

Platform

1. Disposal cost $150,744
2. Platform additions allocated
   to one well (based on total
   6 wells on platform)   5,000
3. Cost of rig downtime due to
   weather (same as above except
   daily operating cost is
   $22,000 on stand-by)   154,000
                                  Total $309,744

A similar cost estimate was recently made for an 18,000 foot
well on Georges Bank which would take approximately 150 days to
drill. These figures exclude the cost of handling and discharge once
the waste materials reach shore. Due to the rougher environment,
barges with their restricted sea state capabilities probably would
not be used. Specially modified work boats capable of offloading in
seas up to 10 feet would be used.

1. Total volume cuttings   521 yd^3

2. Total weight cuttings   439 tons

3. Mud - volume   4,307 barrels
        weight   904 tons

4. Vessel costs
   a) basic vessel $3,800/day
   b) fuel   1,200/day
   c) vessel modification   1,500/day
                              Total $6,500/day

5. Drilling vessel daily rate $ 70,000

6. Standby days due to weather   16 days

7. Estimate Costs

   Vessels   150 days x $13,000   1,950,000
   Vessels Standby
      16 days x $13,000   208,000
   Drill rig standby
      16 days x $70,000   1,120,000
   Downspout   40,000
                              Total $3,318,000

If the tugs and barges were used, costs could soar to as much as $6.2 million per well. Based on these estimates, if 100 wells were drilled in an operating environment similar to Georges Bank, an additional $300,000,000 could be added to the drilling costs. This is a major cost and very careful analysis should be made in the future to determine if the environmental hazards truly justify this added expense to the industry and the consuming public.

Environmental Studies

There has been considerable research activity into the environmental fate and effects of drill muds and cuttings during the past half dozen years. Government research has been conducted by the Environmental Protection Agency and the Department of Energy.

The petroleum industry has been involved in research along two fronts. First are offshore studies that have been linked to lease stipulations and NPDES permits (Table 2). These have addressed both fate and effects. Marine toxicology has been the focal point of two American Petroleum Institute (API) programs which have been addressing warm and cold water species.

The early literature on the toxicity of drilling fluids primarily referred to freshwater species and a few marine invertebrates. More recent data resulting from two API projects (Appendices 1 and 2) and the ARCO Cook Inlet study (Appendix 3), cover a broad spectrum of estuarine and marine organisms. In acute bioassays (96 hr. LC50), most organisms require in excess of 10,000 parts per million to kill 50 percent. Sublethal responses can be detected down to 1,000 parts per million in some species.

In order to better understand the potential hazards of ocean discharge of these materials, field studies have been conducted to study both the fate and effects (Table 2). As can be seen in Appendices 4, 5, and 6, results from the Tanner Bank and Gulf of Mexico discharges studies show the extremely rapid dilution that occurs both during normal drilling discharges and bulk mud releases.

In most cases, dilution of the whole mud is in the range of 100,000:1 within the first 200 meters. Concentrations that would occur in a drill mud plume are far below those necessary to create toxicity problems for water column organisms.

TABLE 2

Petroleum Industry
Monitoring Studies:
Drill Muds and Cuttings

Company	# Studies	Approx. Cost ($ x 1000)[1]	Area
Shell	1	400	S. Calif.
ARCO	1	500	Cook Inlet, AK
Exxon/OOC	1	600	Texas
Exxon[2]	1	820	Baltimore Canyon
Sohio	1	750	Beaufort Sea
Mobil	2	525	Gulf of Mexico
Union	2	275	Gulf of Mexico
G-1 Cost Well[3]	1	75	Georges Bank
Conoco		1,000	Gulf of Mexico
		$4,945	

(1) All costs are approximate.
(2) Exxon served as primary operator, representing all companies
       with Baltimore Canyon leases.
(3) A multi-company sponsored project.

Near field benthic organisms may be effected for short periods of time due to the mud and cuttings settlement near the drilling rig. Community effects would probably be due more to sediment size effects than toxicity. Recovery is thought to be rapid in offshore areas and probably explains the lack of detectable impacts in previous OCS studies.

With the present state of the knowledge on fate and effects of drill muds and cuttings, the petroleum industry does not believe cessation of discharge is necessary. Site specific consideration of discharge procedures in biologically sensitive areas should insure that the marine environment will not be damaged.

Dr. Ray is Staff Specialist Marine Biology in the Environmental Affairs Department, Shell Oil, Houston, Texas. Prior to joining Shell, he earned his MS and Ph.D degrees from Texas A & M University in Marine Biology and Biological Oceanography. For the past five years, Dr. Ray has been involved in various industry research programs to better understand the environmental fate and effect of drill muds and cuttings. He is currently the chairman of the API Drilling Fluids Bioresearch Task Force, and also a member on several industry advisory groups involved with drilling fluids research.

Appendix 1

API DRILLING FLUIDS RESEARCH
TEXAS A&M UNIVERSITY

Acute bioassays and chromium bioavailability studies performed
to date with four used drilling muds. Positions marked with an
X represent bioassays that are proposed to provide a broad overview of comparative toxicity of used drilling muds.

Species	Bioassay Type					Bioavailability	
	MAF[a]	FMAF	LSP	SP	SPP	Cr	Other Metal
Mollusca							
Mercenaria campechiensis	1[b]			1			
Donax variabilis - juv.	1	1	1				
adult	1X	1	1X		X	X	X
Aequipecten amplicostatus			1				
Rangia cuneata				1		1X	X
Crassostrea gigas - spat					2,3,4	3,4X	X
Annelida							
Neanthes arenaceodentata							
juv.	1	1	1				
adult	1X	1	1X		X	X	
Ctenodrilus serratus	1	1					
Dinophilus sp.	1						
Ophryotrocha labronica	1,3X						
Arthropoda							
Penaeus spp. PL	X		X	1	X	X	X
juv.	1						
adult				1		1	
Portunis spinicarpus				1		1	
Mysidopsis almyra juv. 1d	1.sX	1	1				
3d	1						
7d	1						
14d	1						
Palaeomonetes pugio PL	1						
juv.	1						
adult	1X						
Clibanarius vittatus	1.2X						
Callinectes sapidus juv.	2						
Echinodermata							
Mellita quinquiesperforata			1				
Vertebrata							
Fundulus heterolcitus							
emb.	1X		1X				
fry						1	
adult	2					X	X
Menidia berryllina	1	1					
Cyprinodon variegatus	1						

a. MAF, mud aqueous fraction; FMAF, filtered mud aqueous fraction, LSP, laye solid
phase; SP, suspended phase; SPP, suspended particulate phase.

b. 1. chrome lignosulfonate mud; 2, high weight lignosulfonate mud; 3, mid-weight
lignosulfonate mud; 4, spud mud.

Appendix 2

API DRILLING FLUIDS RESEARCH
BRUNSWICK, MAINE

	Suspended Particulate Phase[1]	Liquid Phase[2]	Solid Phase[3]	Whole Mud[4]
ANNELIDA				
Nereis virens[5]				
glucose-6-phosphate dehydrogenase enzyme activity (G6PdH)				
96 hrs.	+			
aspartate aminotransferase enzyme activity (AAT)				
96 hrs.	+			
96 hr. LC50	+		+	+
Cadmium, Chromium uptake	+		+	
availability of Cd, Cr	+			
96 hr. LC50 CrCl$_3$				
depuration of Cd and Cr	+			
MOLLUSCA				
Illyanassa obsolata (snail)				
chemoreception				
Littorina littorea (periwinkle)				
144 hr. LC50	+			
Macoma balthica				
96 hr. LC50	+	+		+
Burrowing behavior				+
Mytilus edulus				
APT enzyme 96 hr.	+			
96 hr. LC50	+			
Cr uptake 96 hr. exp.	+			
Sublethal O:N ratio 96 hr.	+			
Respiration rate 96 hr.	+			
NH$_4$ excretion rates 96 hr.	+			
Filtration rates 96 hr.	+			
Cd uptake	+			
10-day flo-thru growth rate study				+

Appendix 2 (continued)

	Suspended Particulate Phase[1]	Liquid Phase[2]	Solid Phase[3]	Whole Mud[4]
CRUSTACEA				
Gammarus sp. (amphipod)				
96 hr. LC50 (male and female)	+	+		+
Carcinus maenus (green crab)				
chemoreception-feeding response	+			
G6PdH activity 96 hr.	+			
96 hr. LC50	+			
Crangon septemspinosa (sand shrimp)				
96 hr. LC50	+			
24 hr G6PdH enzyme activity	+			
96 hr. G6PdH activity	+			
Cd and Cr uptake 96 hr.	+			
Heavy metal depuration	+			
Pandalus borealis (shrimp)				
96 hr. LC50 larvae	+			
Nomarus americanus (lobster)				
Stage 4 and 5 larvae 96 hr. LC50	+			
VERTEBRATA (FISH)				
Fundulus heteroclitus (killifish)				
96 hr. LC50	+			
enzyme test (96 hr.)	+			
Alosa pseudoharrengus (alewive)				
egg exposure (24-36 hrs.)				

FIELD STUDY

Settling trays with varying mixtures of drill mud have been on the bottom of Bethel Bay. These will be studied for effects of drill mud on _in situ_ benthic recruitment.

[1] Typical of liquid phase and light suspended particulates found in plume.
[2] Liquid phase with all particulates removed.
[3] Heavy suspended particulates that settle out rapidly.
[4] Field mud as discharged with liquid and solid phase.
[5] All species have been tested with "types" of muds typically used off New England coast.

Appendix 3

ATLANTIC RICHFIELD STUDY - LOWER COOK INLET
DAMES AND MOORE, 1978

	Test Duration	Aeration	Stirring	LC50 (ppm by volume)
SHRIMP				
Pandalus hypsinotus	96 hr.	Yes	Minimal	>100,000
(Coonstripe shrimp)	96	Yes	Minimal	32,000
	96	yes	Minimal	>100,000
	96	Yes	Minimal	86,000
	96	Yes	Paddles	44,000
	96	Yes	Minimal	150,000
	48	Yes	Paddles	5C,000-<100,000
	48	Yes	Paddles	>100,000
AMPHIPODS				
Anisogammarus				
confervicolus	48	Yes	Minimal	10,000 - 50,000
	48	None	Minimal	10,000 - 50,000
	96	Yes	Dynaflo	>70,000
	96	Yes	Minimal	>200,000
MYSIDS				
Neomysis integer	48	None	Minimal	100,000 - 150,000
	96	Yes	Paddles	10,000 - 50,000
	96	Yes	Minimal	100,000 - 125,000
	48	Yes	Minimal	74,000
	48	Yes	Minimal	>100,000
ISOPODS				
Gnorimosphaeruma				
oregonensis	96	Yes	Dynaflo	>70,000
BRINE SHRIMP				
Artemia salina	48	None	Minimal	>100,000
	48	None	Minimal	>100,000
MOLLUSC				
Modiolus modiolus	326	Yes	Minimal	>300,000
FISH				
Oncorhynchus				
gorbuscha (Fry)	96	Yes	Dynaflo	19,000
(Pink salmon)	96	Yes	Paddles	3,000
	96	Yes	Minimum	29,000
Leptocottus armatus				
(Staghorn sculpin)	48	Yes	Paddles	100,000 < 200,000

In Situ Live Box Tests - 100 and 200 meters downstream - three depths
  96 hr. exposure 100% survival - all species - maximum discharge rate
    Pandalus hypsinotus - Coonstripe shrimp
    Oncornynchus gorbuscha - Pink salmon fry
    Hermit crab

91

Appendix 4

[1]Tanner Bank Mud and Cuttings Study

[2]Maximum Mud Dump   750 bbl/hr

	TSS	Ba	Cr
Whole Mud	250,000 mg/l	13,000 mg/l	286 mg/l
0-3 meters	328 mg/l	12,700 µg/l	917 µg/l
74 meters	25.2 mg/l	575 µg/l	13.5 µg/l
500 meters	4.04 mg/l	146 µg/l	16.4 µg/l
625 meters	1.10 mg/l	47.2 µg/l	0.528 µg/l
800 meters	4.73 mg/l	111 µg/l	7.37 µg/l
1,000 meters	0.563 mg/l	26.2 µg/l	0.916 µg/l
Control	0.814 mg/l	21.9 µg/l	0.481 µg/l

Typical cuttings discharge with Associated Mud

	TSS	Ba	Cr
Whole Mud	250,000 mg/l	13,000 mg/l	286 mg/l
0-3 meters	7.73 mg/l	36.2 µg/l	3.4 µg/l
50 meters	2.16 mg/l	28.3 µg/l	1.13 µg/l
113 meters	1.56 mg/l	12.9 µg/l	0.549 µg/l
200 meters	2.96 mg/l	21.8 µg/l	0.706 µg/l
350 meters	2.35 mg/l	13.7 µg/l	0.845 µg/l
Avg. Control	0.8-1.2 mg/l		

[1]Ecomar, 1978

[2]Samples taken from area of maximum density in the discharge plume.

Appendix 5

TANNER BANK STUDY
EFFECT OF DRILLING FLUID DISCHARGE ON WATER QUALITY

Run	Distance From Source Meters	Suspended Solids mg/l	Barium* mg/l	Chromium mg/l	Lead mg/l	Transmittance %	pH	Dissolved Oxygen mg/l	Salinity ppt	Temp °C
Drilling Fluid	--	250,000	14,000	302	26.5	--	--	--	--	--
A Vol = 5 bbl Rate = 10 bbl/hr	<1	499	23.6	0.824	0.038	--	--	--	--	--
	105	5.2	0.103	0.004	0.0004	49.1	8.39	10.05	32.0	13.8
	155	2.03	0.047	0.008	0.004	62.8	8.4	10.11	32.8	14.0
	450	1.79	0.038	0.008	0.005	77.1	8.43	10.01	33.6	13.7
	Control	1.54	0.013	0.004	0.034	83.44	8.45	10.06	33.4	13.7
G Vol = 125 bbl Rate = 750 bbl/hr	<1	328	12.7	0.917	0.04	--	--	--	--	--
	74	25.2	0.575	0.013	0.003	0.0	8.44	9.8	32.8	14.2
	500	4.04	0.146	0.016	0.0009	19.3	8.48	9.99	33.6	14.2
	625	1.10	0.047	0.0005	0.0005	80.8	8.46	10.02	34.0	14.6
	800	4.73	0.111	0.0007	0.0044	23.7	8.46	10.06	33.8	13.9
	1000	0.563	0.026	0.0009	0.0001	10.9	8.47	10.00	34.1	13.9
	Control	0.814	0.022	0.0005	0.0002	94.8	8.44	10.03	34	13.9

*Present as $BaSO_4$

Appendix 6

OFFSHORE OPERATORS COMMITTEE/EXXON GULF OF MEXICO
MAXIMUM MUD DISCHARGE STUDY

SUSPENDED SOLIDS CONCENTRATION AND TRANSMITTANCE VS. DISTANCE
DURING HIGH RATE DISCHARGES

	275 Barrel/hour	250 Barrel Discharged
Distance from Source, Meters	Solids Concentration, Mg/l*	Transmittance, %*
0	1,426,675	-
6	14,756	0
65	34	2
138	9	56
250	7	48
361	16	37
625	1	72
Background	1	83

	1000 Barrel/hour	389 Barrel Discharged
Distance From Source, Meters	Solids Concentration, Mg/l*	Transmittance, %*
0	1,426,675	-
60	32	0
152	51	2
376	24	15
498	9	25
777	4	30
1470	2	82
1564	1	82
Background	1	86

*Maximum concentration and minimum transmittance measured at noted distance.

SUBSEA PRODUCTION SYSTEMS

J. Preston Mason

Background

Subsea production systems are made up of wells completed on the sea floor and connected by flowlines and controls back to a surface facility. As you will see, subsea systems will tie together many components including drilling, pipelines and flowlines and offshore storage and floating production facilities.

The first known underwater completion (UWC) on the North American continent was made in 1943, in 35 feet of water in the Canadian waters of Lake Erie. Since then, more than 300 UWC's have been made in Lake Erie, and this represents the largest concentration of UWC's in the world. These have all been relatively low pressure (less than 2,000 psi) gas wells in shallow water (less than 85 feet). The wells are equipped with simple land type Xmas trees which require divers to install, connect flowlines and operate the valves.

Development of deep water subsea wellhead equipment and completion technology was not seriously undertaken for the open sea until the early 1950's. During the mid 1950's, R&D work was begun to develop remotely operated equipment and techniques for sea floor well completions.

Figure 1 presents a plot of the number of sea floor wells versus years from 1960. The number of sea floor completions was less than 10 per year during the sixty's and early seventies. Many of the early trees were diver installed and operated by hydraulic remote controls. Many were experimental. By the mid 1960's and later, more of the wells were equipped for electrohydraulic control and many were considered commercial as opposed to experimental. In the mid 1970's the number of sea floor wells being installed increased. In 1979, 21 wells were completed on the seafloor by nine oil operators. There are now about 59 subsea trees on order indicating that the number of sea floor wells will continue to increase.

Another interesting trend is the number of companies which have had experience with sea floor wells. Figure 2 shows a plot of the cumulative number of companies that have now installed at least one sea floor well. Corresponding to this trend is the list of manufacturers supplying subsea trees. They are:

1.	Cameron	5.	National
2.	FMC/OCT	6.	Regan
3.	Lockheed	7.	Vetco
4.	Envoy	8.	WKM

95

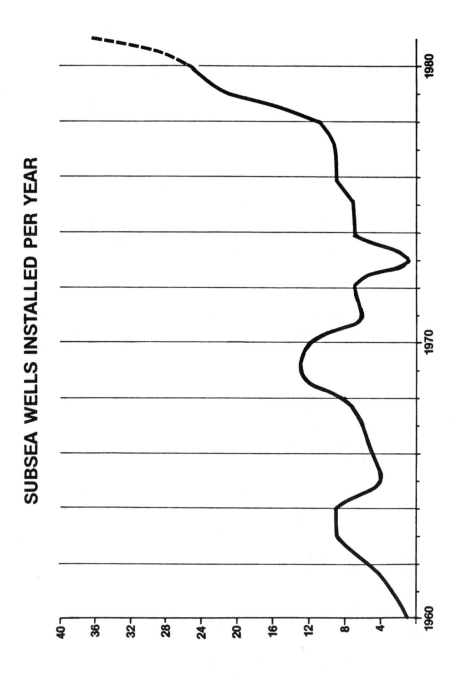

Figure 1

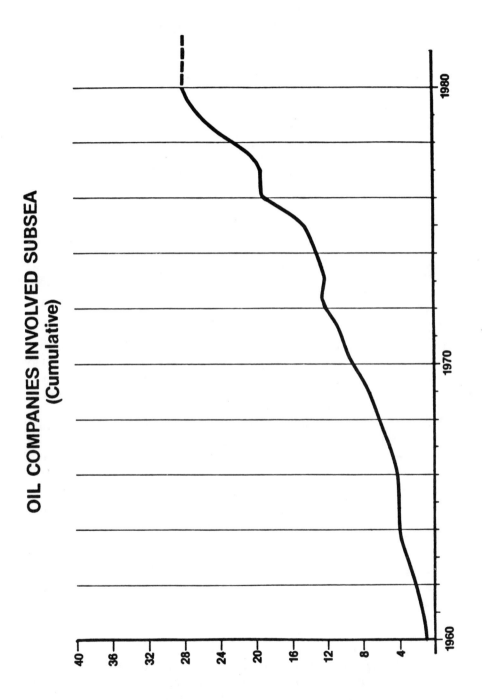

Figure 2

The points of this background are:

-   Evolution of sea floor well technology has
    been a slow process. As will be seen in
    this paper, industry has moved from adapt-
    ing land trees to the point where modern,
    remotely installed and operated systems are
    available.

-   Many oil companies and manufacturers have
    invested a lot of money and manpower to
    develop the technology to be ready when
    economics justify moving into deepwater
    or hostile environments.

Available Production Systems

Subsea production systems range from single wells to total
production systems. A single well completed on the sea floor is
referred to as a "satellite" well and is connected by flowlines to a
remote central facility. Most of the subsea completions to date have
been satellite wells connected to platforms in shallow water. These
wells have been used to drain outlying areas of fields which could
not be drilled from the central platform. Total production systems
include wells, sea floor gathering, risers connecting the sea floor
to the surface, production facilities, storage and transport.

A schematic subsea completion is shown in on the left-hand side
of Figure 3 to define some of the terms used in this presentation.
The downhole completion below the mud line consists of various sized
casing required to maintain structural integrity of the well and
allow well control during drilling. A tubing string extends from the
wellhead to the producing formations. The wellhead supports the
casing and tubing at the mud line. Control valves start with the
downhole safety valve used to shut-in the flow in an emergency. The
master valves are used to secure the well in normal operations after
flow is stopped by the wing valves. A crossover valve is provided to
allow connection of the tubing-casing annulus with the flowline.
Swab valves allow vertical entry into the well from a drilling rig or
service vessel located overhead.

In some cases, two flowlines are used as shown on the right-
hand side of Figure 3. One line is connected directly to the
production tubing and the other to the annulus with a crossover
between the flowlines. This arrangement allows monitoring of the
annulus pressure without stopping production, provides a circulation
path and parallel flowlines for flexibility in operation, and
redundancy in case of damage to one line.

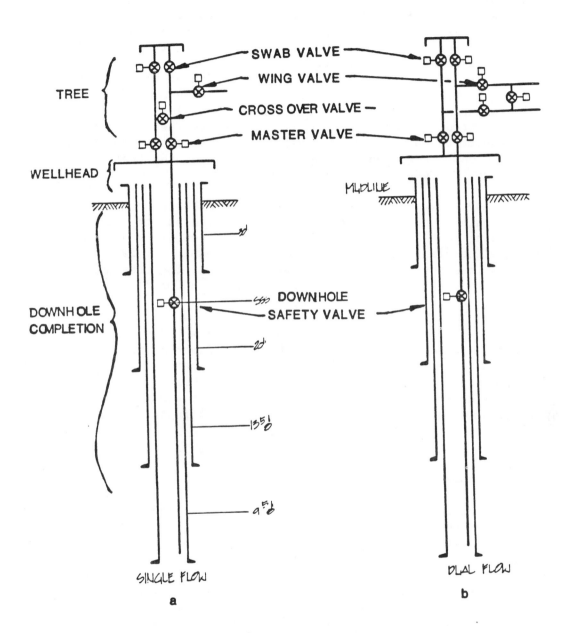

SWAB VALVE

WING VALVE

CROSS OVER VALVE

MASTER VALVE

TREE

WELLHEAD

MUDLINE

DOWNHOLE
SAFETY VALVE

DOWNHOLE
COMPLETION

SINGLE FLOW

a

DUAL FLOW

b

Seafloor Completion Flow Schematics

Figure 3

Satellite Wells

An example of a satellite well system with dual flowlines is shown in Figure 4. This is a single tree which consists of a block containing the master and swab valves. This valve block is very similar to those used for trees on platforms. It is attached to the wellhead by a hydraulically actuated connector. Wing valves and the crossover valves are separate and are located adjacent to the valve block in the flowlines. All valves are hydraulically operated by controls from the surface. They are designed to fail to the safe position if hydraulic control pressure is removed. The flowlines on the tree connect to the sea floor flowlines at the edge of the guide base near the mud line. Remote control equipment on the tree is located in the tree cap for ease of maintenance.

This tree is designed to be installed on guidelines from a surface rig using remotely controlled running tools and to operate in the surrounding sea water. Installation of the controls and connection of flowlines is by diver assist. Downhole equipment is maintained by reentry through the top of the tree. The tree and controls are maintained by retrieval to the surface for repair. The tree is thus referred to as a "wet" system.

Many of the subsea trees installed around the world have been of this type. The tree in Figure 4 is marketed by Cameron Iron Works as a "Plain Jane" model. This particular unit fits over a standard 8 foot guidepost spacing, is about 18 feet tall and weighs about 20,000 lbs. Units like this equipped with fully redundant electrohydraulic controls have recently been installed by Mobil Oil in the Beryl Field in the North Sea.

For application in deeper water or more hostile environments, additional capability has been developed.

Flowline connectors and remote control equipment modules and tools have been developed to allow installation of subsea trees without use of divers. In addition, tools which can be pumped down the flowline and into the wells have been developed to permit operation and maintenance of downhole equipment. Through Flowline (TFL) or Pump Down (PDT) tools are now proven for performing essentially all functions that are done by wireline operation in wells located on the surface. Use of TFL requires the use of two flowlines, and all bends in the flowlines must be five foot radius or larger.

"PLAIN JANE" TREE
(No TFL)

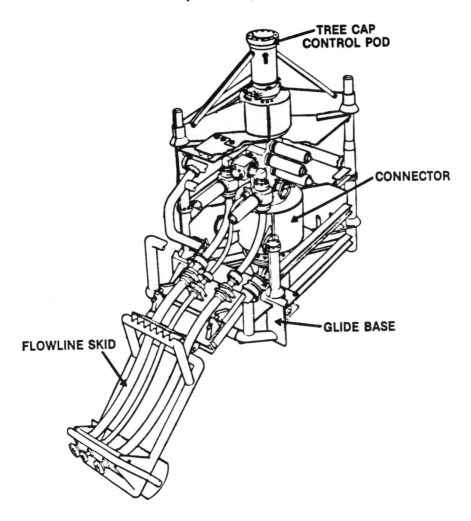

TREE CAP
CONTROL POD

CONNECTOR

GLIDE BASE

FLOWLINE SKID

Figure 4

A subsea tree equipped for TFL maintenance and diverless installation is shown in Figure 5. The tree is equipped with a diverless flowline pull-in and connection system, a diverless control module, and a pop-up buoy system which allows reestablishing guidelines in deep water. The flowlines make a five foot radius loop as they change from vertical to horizontal on the tree. A tree such as this one manufactured by VETCO Offshore could be installed in any depth where wells can be drilled. A tree like this would be run on standard guidelines or by use of guidelineless systems depending on water depth. The tree would be about 25 feet tall and weigh about 30,000 lbs. There are now several manufacturers who can provide this system which can be installed without divers and at great depth.

Another approach to maintenance of subsea wells is to install the sea floor equipment in a chamber. Air is maintained in the chamber at one atmosphere pressure to allow men to enter the chamber and perform installation and maintenance tasks. The concept is to allow use of land equipment and procedures subsea.

This "dry" system concept was developed by Lockheed Petroleum Services in conjunction with Amoco and more recently with Shell Oil. In the Lockheed system shown in Figure 6, the well is drilled and then the chamber is installed. Men then go to the chamber and assemble the tree and assist in flowline pull-in connection.

Men can be transferred to the well in a tethered bell type transfer vehicle or a submersible. The tethered bell unit is shown mated to a wellhead chamber in Figure 7. As shown, life support and power would be supplied down the umbilical.

Several wells have been installed in the Gulf of Mexico and offshore Brazil inside Lockheed well chambers. In addition, the pipeline connections to the platform in the Thistle Field in the North Sea were made inside Lockheed chambers. Cameron Iron Works has also recently developed a dry system. The technology of this approach is well proven. The system could be used to any depth, limited only by economics. Presently available equipment is rated to water depths of 1,500 feet. Deeper water units have been designed.

A recent development is a subsea tree which is installed below the mud line. As shown in Figure 8, the "tree" connects to a wellhead located about 70 feet below the mud line. A master valve block is located above the connector. A diverter is shown to allow use of TFL tools. Only the swab valves, tree cap and flowline

SATELLITE TREE
(with TFL)

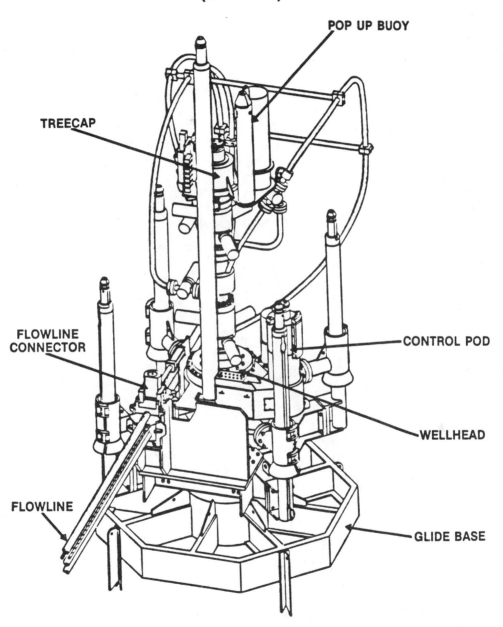

POP UP BUOY

TREECAP

FLOWLINE
CONNECTOR

CONTROL POD

WELLHEAD

FLOWLINE

GLIDE BASE

Figure 5

ONE ATMOSPHERE SUBSEA WELL CHAMBER

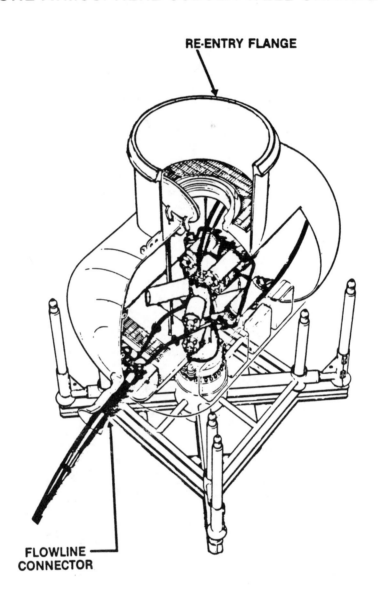

Figure 6

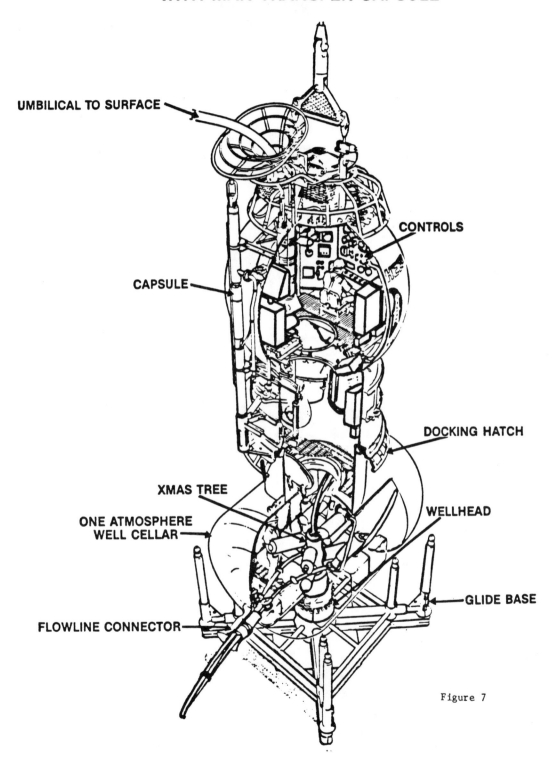

WELL CHAMBER
WITH MAN TRANSFER CAPSULE

UMBILICAL TO SURFACE

CONTROLS

CAPSULE

DOCKING HATCH

XMAS TREE

ONE ATMOSPHERE
WELL CELLAR

WELLHEAD

GLIDE BASE

FLOWLINE CONNECTOR

Figure 7

INSERT TREE
(with TFL)

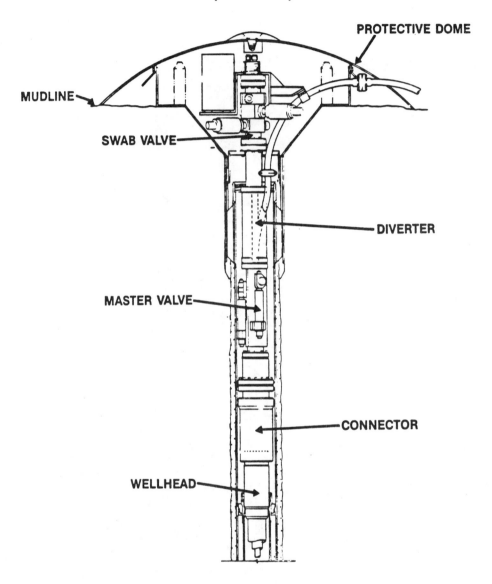

Figure 8

connector extend above the sea floor. The top of the tree is only about 10 feet above the mud line compared to about 35 feet for a conventional TFL tree.

Cameron Iron Works developed special slimline hydraulic actuated connectors and ball valves to allow their "Caisson" tree to fit inside a 30-inch casing. A prototype of that unit has been built and tested. The first tree for offshore installation is undergoing shop testing. This concept should be ready for use where needed within the next few years.

Semisubmersible Production System

The system shown in Figure 9 consists of satellite wells connected by flowlines to a sea floor base located directly under the semisubmersible. The semisubmersible will be a converted drilling rig kept on station by a conventional chain and wire rope spread mooring. Anchors will probably be drilled in to allow the vessel to stay on location during very severe weather.

The production riser will carry fluids between the sea floor manifold and the semi-submersible vessel. All sea floor equipment is controlled from the surface. Process and injection facilities will be located on the deck of the semisubmersible. Separated oil will be flowed down the riser, through a sea floor pipeline, up a single point mooring and to a shuttle tanker for periodic transport to market.

For small reservoirs, the system shown would be used. This would include satellite wells, probably completed as part of the exploration, field delineation drilling. The wells would be brought directly up the riser to the vessel with no sea floor manifold. The production rate would be smaller so that only the limited oil storage available on the semisubmersible would be justified. Production would be interrupted when the shuttle tanker was out of the berth for more than a few hours. The volume of gas would be small and, unless the field was near existing pipelines, the only economic way to handle the gas would be to flare it.

For larger reservoirs, the number of wells would increase and it would be necessary to manifold the flow on the sea floor to keep the flowlines in the riser to a practical number. Oil storage would be economically justified and would probably be included by using a permanently moored tanker. Oil will be periodically transferred to a shuttle tanker for transport to market. The shuttle tanker would be docked alongside or tandem moored to the storage unit. In addition, gas could be recompressed and reinjected underground, or flowed by pipelines to shore.

108

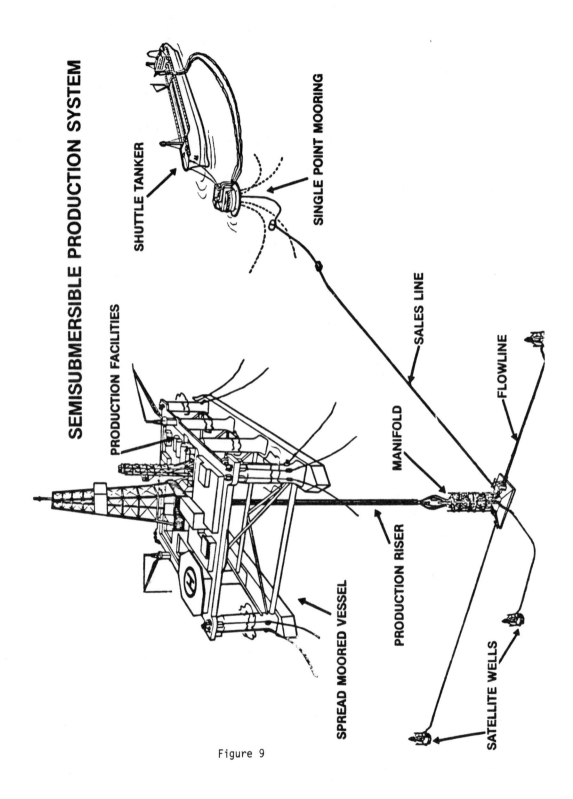

SEMISUBMERSIBLE PRODUCTION SYSTEM

SHUTTLE TANKER

SINGLE POINT MOORING

PRODUCTION FACILITIES

SALES LINE

FLOWLINE

MANIFOLD

SPREAD MOORED VESSEL

PRODUCTION RISER

SATELLITE WELLS

Figure 9

This type of system was the first commercial use of subsea wells to develop an entire field. Hamilton Brothers used the system to develop the Argyll Field in the North Sea. Production started in 1975 and the system has operated satisfactorily since that time.

A variation of this system is shown in Figure 10. A multiwell template is located below the submersible. Wells are drilled through the template. Subsea trees are installed on the wells and connected to piping on the template which connects to the production riser. Satellite wells can also be drilled and connected to the template as shown. The particular system depicted in Figure 10 is now being installed in offshore Brazil.

One Atmospheric System

The one atmospheric chamber production system is shown in Figure 11. Enclosed individual wells are connected by flowlines to a manifold center which is enclosed. The manifolds are connected by pipelines to the base of a single point mooring production riser which transports the fluids between the sea floor and production facilities mounted on a tanker. All subsea equipment would be controlled from the surface. On board the tanker, oil would be separated, stabilized and stored. The oil would then be tranported to market by shuttle tankers which dock periodically with the storage tankers. Gas would be recompressed and injected underground. Produced water would be treated and disposed or combined with treated sea water and injected to maintain reservoir pressure.

If the field is near a market, the oil and gas would be transported by pipeline to shore. In that case, storage in the field would not be required and another type of vessel other than a tanker might be chosen to support process facilities.

Sea floor equipment would be done by men working in the chamber. The men and materials would be transported to site in a tethered manned transfer bell. The wells can be maintained by TFL servicing or by vertical reentry.

Several enclosed sea floor wells and a prototype enclosed manifold have been installed during the last few years in the Gulf of Mexico by Shell Oil. The manifold has been operating for about two years. The first complete field development system was installed last year in the Garoupa Field offshore Brazil. The system will soon be in operation. Lockheed now operates two service vessels equipped to transport men to the sea floor chamber for maintenance.

TEMPLATE & RISER
BUILT FOR PETROBRAS ENCHOVA FIELD

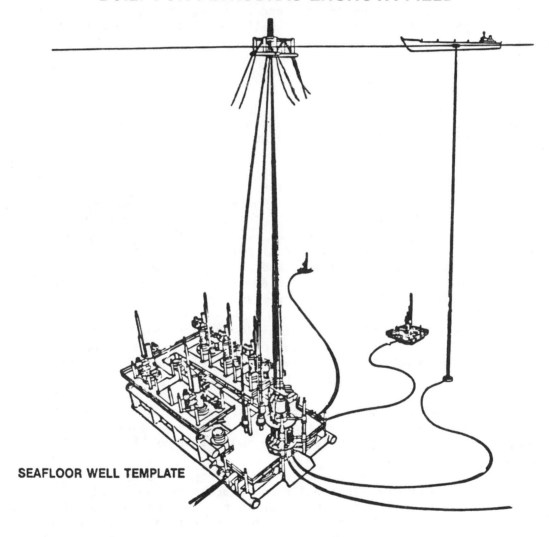

SEAFLOOR WELL TEMPLATE

Figure 10

111

ONE ATMOSPHERE CHAMBER PRODUCTION SYSTEM

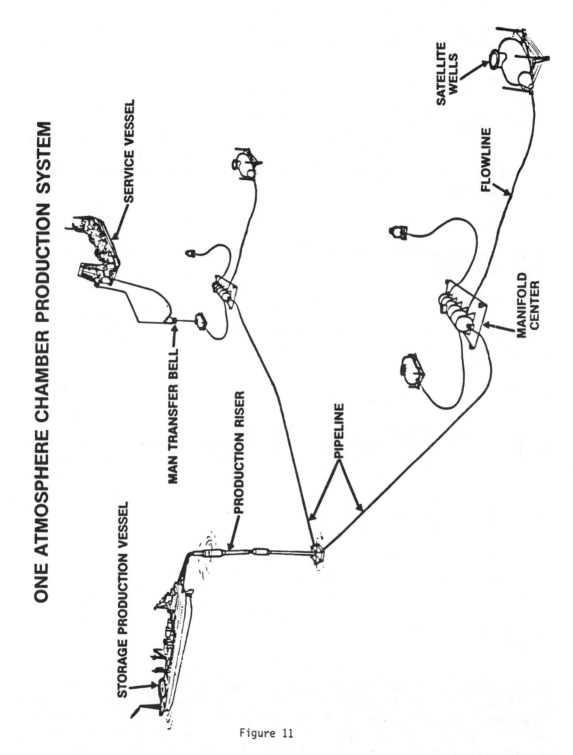

Figure 11

Subsea Atmospheric Systems (SAS)

The production system shown in Figure 12 is a hybrid system developed by Mobil Oil. The wells are drilled through a base template and completed with special wet subsea trees. Those trees are connected to manifolds and remote controls housed in a large chamber. The atmosphere in the lower manifold section of the chamber is inert gas at one atmosphere pressure. A breathable air is maintained at one atmosphere in the upper control and entry section. Men are transported to the chamber in a tethered bell or submarine to do maintenance in the manifold and controls. The wells can be maintained by TFL servicing from inside the chamber or by vertical reentry methods.

The manifolds are connected by pipeline to a production riser and production facilities on a surface vessel. Pipes are also provided to supply nitrogen and air to the SAS. Power and communications are supplied by electric cable.

Work to develop this system began in the mid 1960's and led to installation of a prototype of the SAS offshore in the Gulf of Mexico. Tests during 1972-1974 demonstrated the feasibility of the concept for water depths to 1,500 feet.

Subsea Production System (SPS)

The subsea production system developed by Exxon is shown in Figure 13. In this system, wells are drilled through a sea floor template. The wells are completed which special subsea trees which connect to a manifold circling the well bay area. The manifolds are connected to a production riser by pipeline. Production facilities are located onboard a floating production vessel. Seafloor equipment is controlled from the surface.

Wells are maintained by TFL servicing from the surface station. Sea floor equipment is maintained by a special purpose manipulator shown in Figure 14. That manipulator is operated from a surface vessel to land on a track on the sea floor template. The unmanned manipulator is then controlled from the surface to replace control modules or valves.

Development of this system started in 1968 and concluded with operation of a three well prototype in the Gulf of Mexico during 1974 to 1978. The test included the sea floor template, wells, diverless flowline and pipeline connections, and production riser. The maintenance manipulator was also fully tested. That test demonstrated capability of the SPS to water depths of 2,000 feet and beyond.

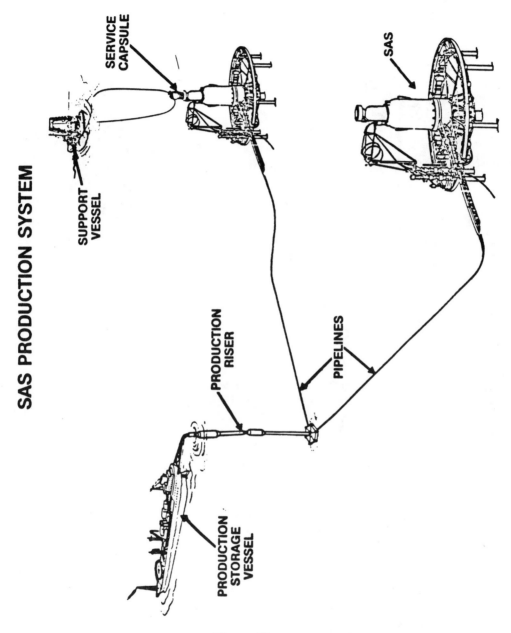

Figure 12

SUBSEA PRODUCTION SYSTEM (SPS)

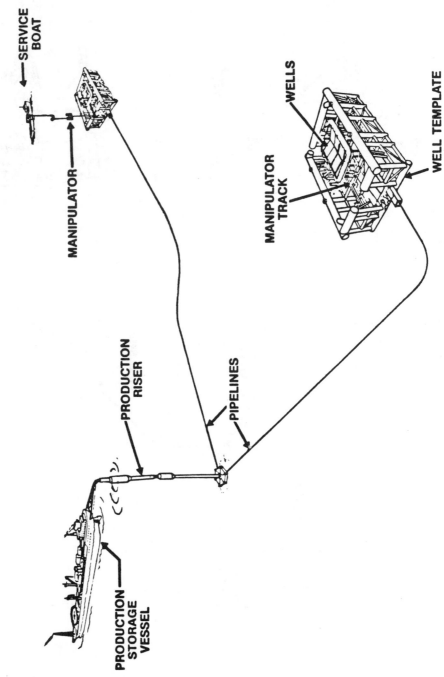

Figure 13

SPS MAINTENANCE MANIPULATOR

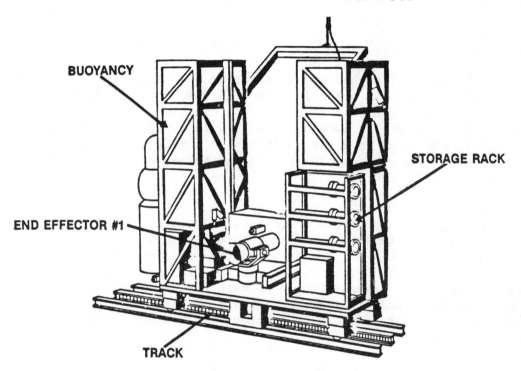

BUOYANCY

STORAGE RACK

END EFFECTOR #1

TRACK

Figure 14

A prototype of a deeper water version of the production riser has been designed and built by Exxon for use in the Santa Barbara Channel.

Technology developed in the SPS program has been combined with satellite well technology by Shell and Exxon for a commercial application in the Cormorant Field in the North Sea. Equipment for that field shown in Figure 15 is now undergoing testing at Bactor, England.

Manned Maintenance Systems

Another general method of installation involves the use of manned manipulator systems as shown in Figure 16 which have been developed by Oceaneering as a backup to diverless running tools. The Jim system is a diving suit worn by a man. Life support is provided through an umbilical from the surface. This unit is lowered from the surface and lands on decks built onto the subsea equipment. The suit is equipped with special tools to allow man to do preplanned jobs. Jim can be used in water depths of 2,000 feet to untangle cables, operate hand valves, or to attach cables. The latest template designed by VETCO for Petrobras (Figure 17) was designed so that Jim can be equipped with special tools for replacing failed valves. All lift capability would be provided from the surface. Since the entire Jim system is transportable by air freight, it can be moved to site and set up in a few days. It can operate off any vessel capable of staying on location. In summary, the Jim system thus provides slightly less capability than a saturation diver but without the long, expensive mobilization costs.

The Wasp is a small swimming submersible worn by a man. This unit has about the same capability as Jim except that it swims rather than walks. Again, several special tasks can be performed by Wasp. The tasks are probably more specialized than with Jim, but less preparation during design is required.

The Arms system is a manned tethered bell equipped with a very sophisticated arm and a grabber arm. The unit is lowered to the sea floor site and landed. The unit is equipped with an anchor and some propulsion to allow the pilot to hover off bottom to do work. The arm includes shoulder, elbow and wrist action and is equipped with position and force feedback. This means that the operator places his hand in a control grip and moves his hand so that he can actually feel the force being exerted by the arm. This capability allows the

MANIFOLD CENTER SYSTEM

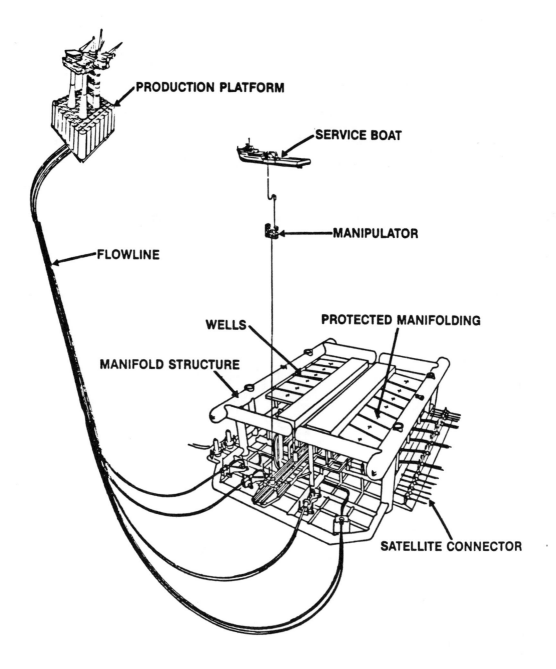

Figure 15

MANNED MAINTENANCE SYSTEM

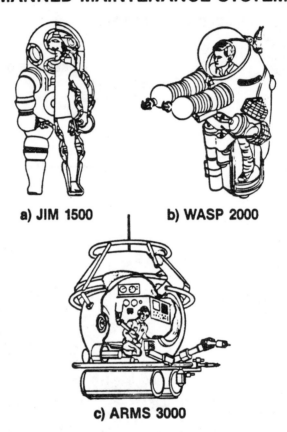

a) JIM 1500   b) WASP 2000

c) ARMS 3000

Figure 16

DRILLING TEMPLATE FOR JIM MAINTENANCE

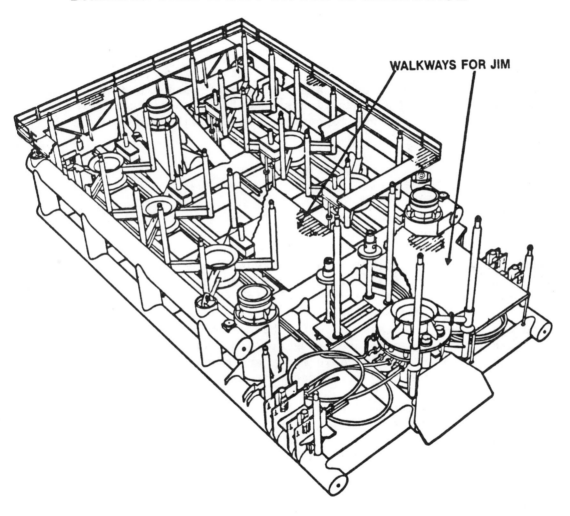

WALKWAYS FOR JIM

Figure 17

Arms to do delicate tasks such as threading a nut on a bolt or installing a hydraulic line. Existing Arms units can operate in water depths up to 3,000 feet.

Discussion

These production systems have several points in common.

Collection of Components

First, all these systems are different. However, on closer examination, each is actually a collection of components which can be configured to design a production system for a specific field. It would be difficult, if not impossible, to say which is best categorically.

Safety

Each system has been designed to be safe, to protect the environment, the reserves, and the investment. During design, engineers have identified problems that could occur and have designed in reliability to avoid problems and back-up or redundant equipment to allow the system to fail safe and to provide recovery from failures. This is referred to as "hazards analysis" in aerospace jargon. In addition, each piece was individualy built and tested before assembly into the system. The systems were then assembled and tested on land and then in the ocean before being applied commercially. The oil companies invested considerable money and manpower in these systems for 10 years before they were ready for use.

The systems are also designed to resist damage. Satellite wells are designed to withstand being hung by fishing gear or anchors from fishing boats or service boats. Template wells are encased in massive structures which will withstand any dragged object and most dropped objects. Flowline connections are designed to withstand pulling of flowlines or pipelines without damage to the well or template structure.

Reliability

Only proven components are used in subsea system, even more so than on land. The equipment is built big and heavy. It is also tested extensively before and during installation. The latest

TABLE 1

Subsea Production System Water Depth Capability

Production System	Water Depth, Feet		
	Installed	Proven	
Satellite Wells	500	4000 +	Fully diverless system installed in Brent Field in North Sea in 1976.
Semisubmersible System	300 +	1000 +	Argyll Field in North Sea, 1975. Buchan Field in early 1980. Designs have been done for application beyond 1,000 foot water depths.
One Atmospheric Chamber	450	3000	Installed in Brazil, in 1977. Designed for 3,000 foot application.
Subsea Atmospheric System (SAS)	225	1500 +	Prototype installed in the U.S. in 1972 using deep water techniques. Prototype chambers certified to 1,500 foot water depth.
Subsea Production System (SPS)	170	2000	Prototype installed in 1974 and operated using diverless deep water procedures and equipment.

technology is used. For example, space age technology is used in valves, flexible joints, and controls system. Each control module on Exxon's SPS has a mean time between failure of in excess of 100 years.

Water Depth Capability

Table 1 is a list of the production systems showing the deepest application to date and the demonstrated water depth. The depth capabilities and dates are compatible with other information presented at this conference. In addition, design of several of these are being extended to 2,500 feet water depths and beyond to keep pace with the current offshore exploration activities.

Mr. Mason is President of Seaflo Systems, Incorporated of Houston, Texas, an organization that he formed in January, 1978. He is responsible for a design and evaluation of various production systems involving seafloor wells and floating production facilities.

From 1968 to 1978, Mr. Mason was involved in the development of Exxon's submerged production system, starting as one of the original design team. He was also involved in application of submerged system technology in the North Sea and in coordination of tests for the system offshore Louisiana.

Mr. Mason holds an MS in Chemical Engineering.

DEEP WATER PIPELINES

J. C. Lochridge

Introduction

During the past several years, the oil and gas industry has focused a
considerable amount of attention on the subject of deep water pipelines. Equipment capable of installing deep water pipelines has been
designed, built and tested, and a number of deep water pipelines have
been installed. The design requirements for such pipelines have been
closely scrutinized as part of deep water pipeline projects and
during joint industry sponsored research programs.

Large diameter pipelines have been laid in the North Sea in water
ranging from 500 to 600 feet deep. At least two test lays using
small diameter pipe in water depths exceeding 1,800 feet have been
made, and a small diameter pipeline (now in service) was installed in
Lake Geneva in 1,100 feet of water. Studies have shown the feasibility of installing large diameter pipe across the Norwegian Trench
(1,100 feet) and intermediate diameter pipe across the Mediterranean
Sea (2,000 to 6,000 feet). At the time of this writing, the so-
called "Trans-Med" project is underway. A pipeline has been
successfully towed across the Norwegian Trench and installed in the
northern North Sea area. A small diameter pipeline was laid from the
Shell "Cognac" platform in 1,000 feet of water in the Gulf of Mexico.

The success of these deep water pipeline projects was dependent
on improved construction equipment and methods, and each project was
supported by large engineering efforts. In anticipation of more such
deep water projects, the industry is continuing to support both
proprietary and joint research studies.

This paper reports on the capabilities of existing pipeline
construction equipment and methods, the development of new equipment
and methods, and the areas of deep water pipelining where research
and development is being pursued.

Pipeline Construction Methods

Although there are number of different methods used to install
marine pipelines, all of the methods can be categorized into two
broad groups. Either the pipeline is welded offshore on location or
it is welded onshore and then transported to the offshore site.

Pipeline Welded Offshore

The method which falls into this category is generally termed the conventional lay method. Pipe joints (40 to 80 feet in length) are welded together on a floating work deck and then lowered to the seabed in a controlled configuration to prevent overstressing.

Pipeline Welded Onshore

Several methods fall into this category and they differ primarily in the manner in which the long pipe strings (a few thousand feet to several miles in length) are transported to the job site. The reel barge, surface tow and bottom tow methods fall into this category.

Conventional Lay Method

This method is the most commonly used for marine pipeline installation. Pipe joints are welded into a continous string on a long, gently curved, production ramp. The anchored lay barge is pulled forward one pipe length as each new joint is added. During pull up, the pipe string passes down the ramp, onto a stinger, and to the ocean floor in an S-curve configuration (Figure 1). Tensioners positioned along the production ramp provide a hold back force which limits the curvature of the pipe string and hence the maximum stress to which the pipe is subjected.

The first marine pipelines were installed in shallow waters using straight stingers and low tension. For deep water, straight stinger lengths become prohibitive, leading to the use of shorter, curved stingers, in combination with higher tension. In order to make optimum use of the stinger length, the pipe exit angle from the lay barge is made as large as possible.

The main limitations of the lay barge method are the downtime incurred by rough sea states, tensioner capacity, the capability of the barge mooring system, and the ability of the pipe to withstand the loads incurred during installation as it passes over the stinger.

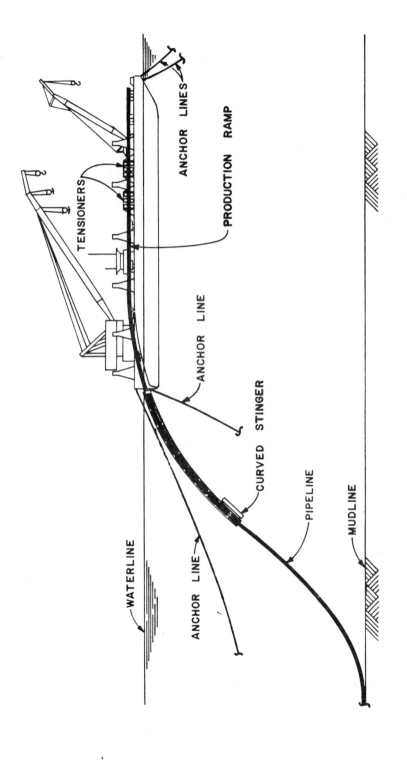

LAYBARGE AND STINGER SYSTEM
FIGURE I

Reel Method

The reel method uses a continuous pipe string assembled onshore and coiled onto a reel. The pipe and reel are placed on a lay vessel and transported to location. At location, the end of the previously laid pipeline is lifted to the surface and welded to the pipe end on the reel. The lay barge is then moved forward, the pipe uncoiled from the reel, straightened, and placed on the ocean floor (Figure 2). This process continues until the reel is empty, whereupon the lay vessel returns to shore to refill the reel.

The reel diameter is such that plastic pipe deformations occur during coiling and uncoiling. In order to avoid excessive pipe flattening during this process, relatively heavy pipe wall thickness is required. At present, 12-inch diameter pipe is the largest to be installed by the reel method, although equipment is available to install 16-inch pipe.

Passage of the pipe into the water can take place at relatively high speed. Most of the job time is for welding onshore, coiling the pipe onto the reel and transporting the pipe to location. Because of the high rate of actual laying, conventional anchor mooring systems are not required, and a form of dynamic positioning is used. Due to the shorter time at sea, the pipelay process is less sensitive to weather and risk of damage during laying.

The main limitations of the reel method are mainly due to the consequences of coiling and uncoiling the pipe. If additional weight is needed to offset pipe buoyancy, it must be provided by increased pipe wall thickness because conventional concrete weight coatings cannot be coiled onto the reel.

Surface Tow Method

The surface tow method requires welding the pipe onshore and towing it to location in long lengths near the water surface. The pipe weight is counteracted by floats spaced along the pipe. At location, the towed length is welded to previously installed lengths on a work deck which resembles a conventional pipelay vessel.

Submerging of the pipe may take place by conventional stinger or by regulating the buoyancy floats (Figure 3). To date this method has not been used extensively for work other than shore approach installations and a small line in Lake Geneva. However, it has been extensively tested in the North Sea.

Shore facilities for pipe make-up require space for about 1,000 feet lengths and protected waters. These lengths are then welded into strings up to several miles in length, and may be pressure

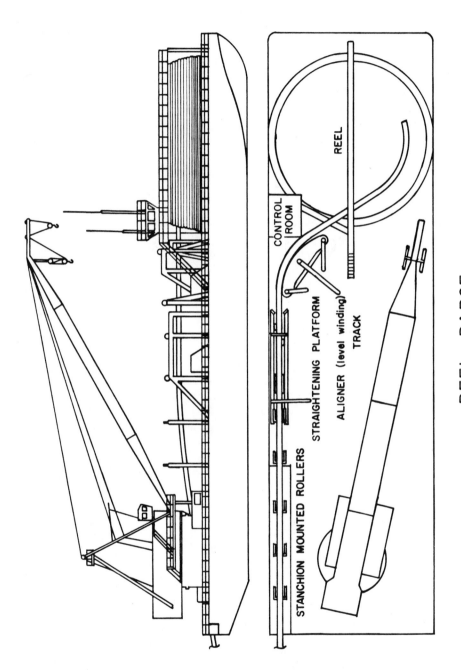

STANCHION MOUNTED ROLLERS

STRAIGHTENING PLATFORM

CONTROL ROOM

REEL

ALIGNER (level winding)

TRACK

REEL BARGE
FIGURE 2

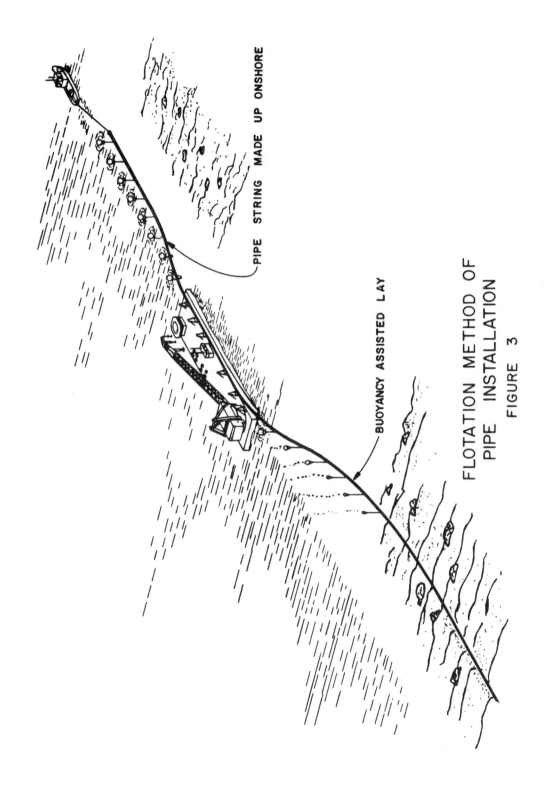

PIPE STRING MADE UP ONSHORE

BUOYANCY ASSISTED LAY

FLOTATION METHOD OF
PIPE INSTALLATION
FIGURE 3

tested before towing. With buoyant floats attached to the pipe, tugs are used to tow the pipe strings to location. Since the flotation method lends itself to a high rate of laying, move up by winching forward on anchor lines and subsequent anchor repositioning may be too slow. Some other form of positioning may be required to develop the full potential of laying speed available with this method. If buoyancy is used during the submerging process, much less tension is required, and perhaps no stinger. Submerging by this process requires reducing the buoyancy with depth and a means of removing the buoyancy floats once the pipe is on the sea floor.

The major limitation of the flotation method is the vulnerability of the pipe to environmental conditions during tow and the difficulty in maneuvering the long pipe strings.

Bottom Tow Method

Like the surface tow method, the bottom tow method utilizes the concept of transporting long strings of pipe made up onshore to an installation site. However, rather than floating the pipe string, it is towed along the sea bottom (Figure 4). Pipe sections are transported to a shore facility close to the installation site, welded and pulled into the water as its length increases.

This method has long been used for pipeline river crossings and for terminals close to shore. The method has also been used for onshore approaches, where the pipe string is made up on a lay barge and pushed or pulled ashore.

When this method is used for the installation of pipelines at large distances from shore, the pipe string is towed by a vessel connected by cable to a pipe pulling head or sled. A bottom tow installation which necessitated crossing the Norwegian Trench was successfully completed in 1977 using this method.

In order to tow a pipe string over great distances and through deep water, several factors must be considered. The required pulling force for only moderate lengths of pipe is large because of the friction between the pipe and the sea bottom. In order to reduce this force the pipe must be as light as possible, but with enough allowance for expected coating losses caused by abrasion and spalling. There must be enough coating left on the pipe after tow to provide the required stability against currents and the corrosion protection should be undamaged. The route over which the pipe string is towed must be carefully selected and surveyed to avoid bottom contours and obstructions which could lead to pipe damage or overstressing.

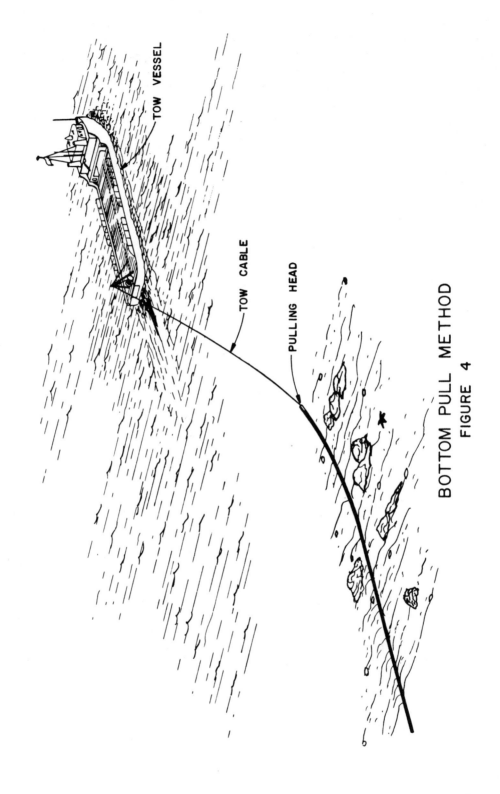

TOW VESSEL

TOW CABLE

PULLING HEAD

BOTTOM PULL METHOD
FIGURE 4

The primary disadvantage of this method is that underwater connections are required if the pipeline is too long to tow in a single length. In this case, several pipe strings would be towed and properly positioned along the pipeline right-of-way. Afterwards, diving teams join the pipe string. The use of this method in constructing long pipelines would greatly be enhanced by the development of underwater diverless connection methods, currently under the study of by several companies.

Pipeline Trenching Methods

Present methods for trenching pipelines consist primarily of using high pressure jets to scour the soil out from under a pipeline which has already been laid, permitting the pipe to settle into the trench thus formed (Figure 5). A sled straddles the pipeline, and is pulled along the pipeline route by a surface barge which contains the jet pumps which provide high pressure water to specially designed nozzles mounted in the sled. The cuttings are removed by air lift, suction dredge pump, or high pressure water eductor.

Until recently, it has been common practice to limit the water depth at which pipe trenching is required to 200 feet and less. It was generally felt that at this depth, sufficient weight coating could be provided to protect the pipe against forces caused by storm waves and currents and against damage caused by fishing activity and anchoring. It has subsequently been recognized that the cost of pipeline repair in water depths greater than 200 feet makes pipeline trenching desirable in areas where damage risks are high.

The construction industry has responded by designing and building equipment capable of trenching pipe in water depths in excess of 600 feet. Some regulatory agencies and owners now require pipelines either to be trenched or otherwise suitably protected against storm wave forces and mechanical damage in areas where there is high risk of damage from trawling or anchors. Common practice calls for lowering the pipeline at least three feet below the existing seabed, and as much as ten feet below seabed close to platforms and in other high risk areas.

The high cost of trenching pipelines has led to a re-examination of the cost/benefit ratio and to the development of more cost effective trenching methods which are discussed later.

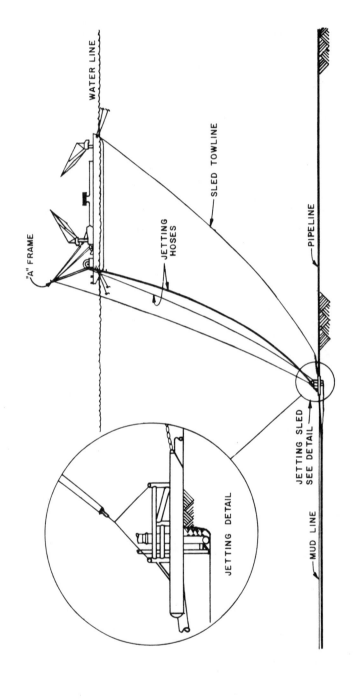

WATER LINE

"A" FRAME

JETTING HOSES

SLED TOWLINE

PIPELINE

JETTING SLED
SEE DETAIL

MUD LINE

JETTING DETAIL

CONVENTIONAL PIPE TRENCHING APPARATUS
FIGURE 5

Deep Water Capabilities Of Existing Equipment

Before describing the deep water capabilities of existing equipment, it is enlightening to review the development of offshore pipelining from its infancy. In the early 1950's marine pipelining was just beginning to leave the Louisiana marshes where it was spawned. Two and one-half decades have seen marine pipelining expand from its humble beginnings to a sophisticated technology. Due to more challenging environments, divergent concepts in equipment design and construction techniques have been developed to meet the needs of the industry.

The more prominent milestones over these last 25 years are shown in Table I. In 1954, Brown & Root's converted dredge barge (Dredge Booth) installed a 10-inch pipeline in what was then considered to be deep water (50 feet). The production line was thought to be highly efficient since the vessel was capable of laying as much as 50 joints of pipe (2,000 feet) in a single 10-hour day. In 1979, both water depth capabilities and production line output have been improved by an order-of-magnitude. This year the Castoro VI is installing pipeline in 2,000 feet of water and the SEMAC-1 has laid 484 joints (19,360 feet) of 18-inch pipe in a single day.

Although a number of different offshore construction methods have been developed over the years, the mainstay of offshore pipeline construction has been the conventional pipelay method. Vessel designs for first, second and third generations pipelay equipment have emerged from barges such as the Dredge Booth. These different categories of pipelay equipment can roughly be described as follows:

a) First generation equipment.

This equipment is similar in design to the BAR-207 (350 ft. x 60 ft. x 22.5 ft., 4,500 ton displacement) constructed in 1958. The vessel quarters 88 men and is equipped for 24-hour production. Pipe handling equipment was automated but no pipeline tensioners were located on the production ramp. Mooring equipment consisted of eight winches driven by 50-hp motors and the drums spooled 3,000 feet of one and one-half inch wire ropes attached to 10,000 pound anchors. Other equipment in this class include the BAR-282, BAR-289 and LB 21.

TABLE 1   <u>MILESTONES IN MARINE PIPELINING</u>

Year	Accomplishment
1954	Dredge Booth lays over 2,000 feet of 10-inch pipe in a day; pipe is laid in up to 50 feet water depth.
1957	M-211 becomes the original first generation laybarge, a flat-deck conversion.
1958	BAR-207 is the first barge initially built for pipe-laying; features include automatic pipe handling and a straight pontoon.
1962	Eight-inch pipe is test laid in 8,300 feet water depth using "breathing" buoys.
1962	The U-303 lays pipe as the first commercial reel barge.
1966	BAR-264, in the first second generation equipment, lays the 16-inch O.D. West Sole Line in the North Sea.
1967	A curved pontoon and tension are used to lay deepwater pipe in the Gulf of Mexico.
1969	Choctaw-I becomes the first semi-submersible laybarge.
1969	LB 22 is the first center ramp laybarge.
1971	Automatic welding used on the Conoco Viking 28-inch line.
1974	The Quille lays 10-inch pipe by surface tow for 1,000 feet depths in Lake Geneva.
1974	Castoro-V lays 10-inch pipe in 1,180 feet depth.

Table 1 (continued)

Year	Accomplishment
1975	Chickasaw lays 10-inch pipe in 1,000 feet depth using the reel method.
1975	BAR-324 lays 279 joints of 32-inch pipe in one day.
1975	Choctaw-II lays 440 joints of 16-inch pipe in 24 hours.
1976	Viking Piper and ETPM 1601 are the first third-generation lay vessels.
1976	ETPM 1601 lays over two miles of 32-inch pipe in one day.
1977	Castoro-V test lays 12 and 16-inch pipe in up to 1,850 ft. depth.
1977	Thirty-six-inch pipe bottom towed to Mobil Statfjord through depths over 1,000 feet.
1979	Castoro-VI laying Strait of Sicily line in depths near 2,000 feet.
1979	SEMAC-1 lays 484 joints of pipe in one day.
1979	McDermott lays Cognac pipeline in 1,000 feet depth.

b) Second generation equipment.

This equipment is typified by the BAR-264 (400 ft. x 100 ft. x 30 ft., 8,300 ton displacement) constructed in 1966. The vessel quarters 250 men and was originally equipped with 250-ton revolving crane so that it could also be used as a derrick barge. The mooring system included eight winches equipped with 3,000 feet of two-inch wire rope and 30,000 pound anchors. Other equipment in this class include LB-22, BAR-323 and BAR-324. The first semisubmersible vessel was also introduced in this class, Santa Fe's Choctaw-I. Automatic pipe welding equipment was also introduced during the development of second generation equipment. The need for the second generation equipment was generated by the discovery of hydrocarbons in the southern North Sea which proved too hostile for the equipment designed for Gulf of Mexico waters.

c) Third generation equipment.

The third generation of pipelay equipment was spawned by the North Sea field discoveries and the need for equipment with deep water capabilities that could operate efficiently in extremely rough sea conditions. Semi-submersible, ship shape as well as the more conventional barge shape hulls have all been used in the design of these pipelay vessels. Many of these vessels are equipped with automatic position control systems and other sophisticated systems. This equipment is characterized by the following:

1. High tension capacity
2. Large pipe storage capacity
3. Advanced mooring systems
4. Large crews
5. Automatic welding
6. Double joining capability
7. Ability to lay pipe in at least 1,000 feet of water

Some of the vessels in this class include the Viking Piper, ETPM 1601, BAR-347, Sea Troll, SEMAC-I and Castoro-VI. The BAR-347 was the culmination of the conventional 'barge shaped' pipelay vessel developed during the first two generations. For comparison purposes, this vessel (650 ft. x 140 ft. x 50 ft., 89,000 ton displacement) can quarter 350 personnel. Its mooring system consists of 12 winches driven by 2,000 hp motors. Each drum winch can spool 10,000 feet of three-inch wire rope and the anchors weigh 60,000 pounds each.

Pipeline Installation Equipment

Table 2 shows some of the deep water work that has been completed in the last six years. In the North Sea, several contractors have laid large diameter pipe in water depths of about 550 feet. In the Gulf of Mexico, a test lay of 12-inch pipe using a reel barge was successfully completed in about 1,000 feet of water. In Lake Geneva, a 10-inch pipeline was installed in 1,100 feet of water, and in the Mediterranean Sea, a test lay in 1,850 feet using 12-inch and 16-inch pipe was successfully completed. Over 15 years ago, French interests successfully made a test lay using buoys in water depths of 8,300 feet. A pipeline has been successfully retrieved from 1,850 feet of water in a test program, and test welds have been made under ambient pressure equivalent to over 1,000 feet of water using welder divers.

In short, it is safe to say that the industry has the capability to lay large diameter pipe in water depths exceeding 1,000 feet. In addition, several industry sponsored programs, supported by dozens of private companies, indicate that pipe of at least 20-inch diameter can be laid in depths of 3,000 feet using existing techniques and knowledge.

TABLE 2   RECENT DEEP WATER PIPELINE INSTALLATIONS

AREA	WATER DEPTH (FT.)	PIPE SIZE (IN.)	CONTRACTORS
North Sea	500	30-36	Brown & Root, ETPM, Santa Fe, Oceanic, Viking
Gulf of Mexico	1,000+	12	Santa Fe (Reel) McDermott (Stinger)
Lake Geneva	1,000	10	O.T.P. (Flotation)
Mediterranean	1,850	16	Saipem (Stinger)

Second Generation Pipelay Vessels

Although the capabilities of this class equipment varies from unit to unit, it is fair to state that most of these vessels were originally equipped to lay large diameter pipe in about 500 to 600 feet of water. As modifications have been made over the years, some of the vessels can lay pipe in water depths of about 1,000 feet under ideal conditions. The main limitations of this equipment are pipe tensioning capacity, mooring system capacity and stinger length. With modifications and improvements in these three basic areas, most of this equipment could be upgraded to lay pipe in 1,000 foot water depths.

It is doubtful that any of this equipment can be economically upgraded to lay pipe in water depths exceeding 1,000 feet unless it is converted to be used as work decks for the surface tow method.

Third Generation Pipelay Vessels

Table 3 lists existing and proposed barges which are claimed to have pipe laying capabilities in excess of 1,000 feet for pipe on the order of 36-inch diameter. These include Brown & Root's BAR-347 and SEMAC-I, ETPM's 1601, Santa Fe's Viking Piper and Apache (reel barge), Sea Troll and the Castoro VI.

TABLE 3   LAY VESSELS
FOR 1,000 FT.+ WATER DEPTH
(Large Diameter Pipe)

BAR-347	Sea Troll
ETPM 1601	SEMAC I
Viking Piper	Castoro VI
Apache (16 inch)	

As mentioned previously, industry sponsored research programs have shown that vessels of this class can lay 20-inch diameter in 3,000 foot water depths. Although some of these vessels may require some upgrading in tension capacity or mooring equipment, it is generally concluded that this class of equipment will perform the majority of the initial deep water pipeline installations.

Reel Vessels

Although not as numerous as the more conventional pipelay vessels, the reel barge offers some advantages in deep water pipeline installations. The method is limited to smaller diameter pipe sizes (16-inch and smaller), but it is less sensitive to rough sea conditions because of the relatively short exposure time during laying. Santa Fe's vessel, the Apache, is rated to 3,000 foot water depths.

Pipe Towing Methods

Pipe towing methods, although not as proven as conventional methods, offer some attractions for deep water applications. The main attraction is less initial capital outlay for the equipment (especially bottom tow methods). The surface tow method offers a natural method of using buoyancy assisted lay procedures since the buoys used for towing could also be used during lay operations. The use of "breather buoys" (which lose buoyancy during their descent) seems a practical means of using this concept. Means of releasing the buoys after the pipeline is on-bottom need to be further developed, but this method will probably be used to a greater extent in the future.

The bottom tow method which is very practical for short deep water installations is more limited for long pipelines. Means of making deep water connections in an economic manner would greatly enhance this method for long deep water pipeline installations. Certainly, the bottom tow method is less sensitive to weather conditions than is the surface tow method.

It is generally concluded that both of the tow methods will be used in deep water pipeline installations. They will require further refinements before they are widely used, and both must pass the test of time before their advantages are proven.

Variations of these methods include near surface tows using spar-like buoys, and near-bottom tows using chains to hold buoyant pipe near the bottom.

Pipeline Trenching Equipment

All modern barges which can trench pipelines in water depths of 600 feet or more are equipped with hose reels. The hoses serve as conduits for the high pressure jet water and the air or water used for removing the cuttings.

One of the limitations of present techniques is the distance behind the barge which the jet sled must be towed in order to preclude an excessive lift from the towing chain. This lift tends to force the sled off the pipeline. This excessive distance precludes efficient diving operations for sled and pipeline inspection and also requires excessive lengths of the very expensive hoses, which are vulnerable to mechanical damage. Notwithstanding these shortcomings, recent studies have shown that it is possible to extend the water capabilities of present equipment to the 1,000 foot water depth range.

Another limitation of the jetting techniques is the high fuel cost associated with operating 40,000 horsepower pump drivers. Fuel costs now are as much as half of the daily cost of a pipe trenching spread.

Deep Water Pipeline Research And Development Efforts

New Construction and Repair Techniques

For years, pipeline installation contractors and pipeline owners have been aware that deep water construction using conventional ideas will require expensive vessels such as third generation equipment. This knowledge has spurred the search for other methods which are more compatible with deep water pipeline requirements. In addition, the industry has been concerned about the repairability of deep water pipeline systems and considerable research has been conducted in the area of deep water diving and repair methods. The discussion which follows briefly describes some of the concepts associated with these developments.

Inclined Ramp

The inclined ramp principle has been known for some time, but has not been used in its purest form for any practical pipelaying. The laying takes place from a lay barge equipped with an inclined ramp. Pipe sections are transported to the lay barge and welding takes place on the ramp. Following the welding, the lay barge moves forward and the pipe is submerged in a J-curve configuration (Figure 6). Tensioners on the inclined ramp are used to support the free hanging pipe weight and to limit the sagbend curvature. The ramp is hinged to the lay barge and will adjust to an angle compatible with water depth and pipe strength. This method eliminates overbend in the pipe and the need for a stinger. However, the production line space is necessarily shorter, and slower laying speeds theoretically result when using conventional welding techniques. To improve laying speed, double-jointing, explosive welding or other single station joining processes could be used. Tension capacity is also limited because of less space in which to place tensioner units.

Pipe Joining

One of the most significant technological breakthroughs which could expedite deep water development would be the development of a practical single-station pipe joining technique. This would allow the use of the inclined ramp method of pipe laying, would extend the capability of conventional lay barges, and would make the flotation and reel barge methods of installation more efficient. Although considerable research has been and is being conducted on single-station joining methods, no one method at present appears on the verge of wide spread acceptability. Welding techniques under study include various automatic GMA systems, narrow-gap, explosive, low vacuum and atmospheric electron beam, variations of laser welding techniques, flash welding, friction, inertia welding, pressure welding, and forging.

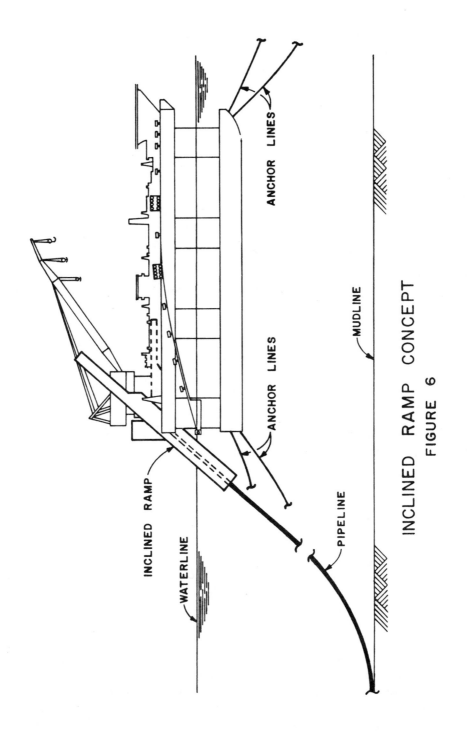

INCLINED RAMP CONCEPT
FIGURE 6

Mechanical Connectors

Despite a wider use of mechanical connectors for risers and tie-ins, the industry has been reluctant to accept them for production joining during installation because of high cost and lack of proven reliability compared to welding.

The apparent cost effectiveness of a simple connector that can be made up on bottom has led about 20 companies to join in an effort to develop new alignment and connector schemes.

Diving and Repair Methods

Another threshold now being crossed for deep water pipe laying activities is the physiological limitation of divers using the most advanced saturated diving techniques. There has been considerable successful research in diving physiology and in welding in water depths between 1,000 and 2,000 feet. While it is possible that diving will extend past 2,000 feet, present forecasts do not predict this occurring in the near future. Therefore, considerable attention has been given to unmanned systems for inspection, repair and other activities required to support offshore pipeline installation and maintenance (Figure 7). It is obvious that an owner will not undertake the financial risk involved in pipeline installation in great water depths unless the capability exists to repair and maintain the pipeline.

There have been several industry sponsored research programs aimed at developing pipeline repair and tie-in techniques for water depths beyond diver capability. Most of these have been reported in technical and trade journals.

Trenching Methods

Pipeline owners and regulatory agencies must determine whether trenching requirements should be extended to greater water depths in spite of the cost and the reduced risk of damage. It is probable that exposure to high currents and storm waves will be minimal, but in selected areas, exposure to trawling activity still exists.

Activity has proceeded on a world wide basis to develop pipe trenching techniques which would be relatively insensitive to water depth, and theoretically more efficient than the proven jetting technique. Most proposed systems utilized mechanical cutters for removing the soil from beneath the pipeline (Figure 8). While the prime movers are located on surface equipment, the hydraulic or electric motors to power the cutters are mounted on a bottom sled which traverses the pipeline. Numerous concepts have been

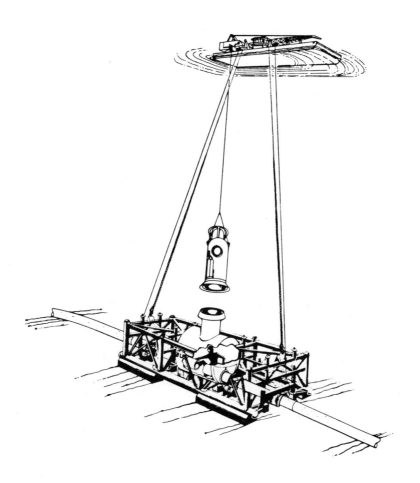

EQUIPMENT FOR DEEPWATER
PIPELINE REPAIR

FIGURE 7

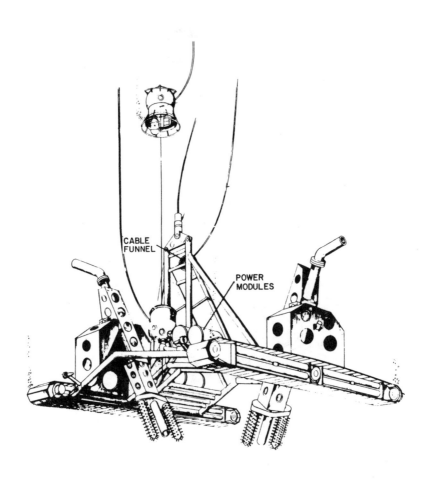

CABLE
FUNNEL

POWER
MODULES

CONCEPTUAL BOTTOM TRENCHING APPARATUS
FIGURE 8

publicized, but as of this date, no system of this type is known to have gained wide acceptance in the industry. Nonetheless, the amount of effort that is going into developing this type of equipment indicates that in the near future a workable prototype will be developed. Pipe trenching plans, for use both before and after pipe lay, have also been proposed.

Deep Water Pipeline Design Considerations

Line Pipe Considerations

Turning now from pipeline installation techniques to the requirements for the pipe itself, the selection of the pipe grade and diameter to wall thickness ratio (D/t) to be used for deep water pipelines requires consideration of the tensile and bending loads and the external hydrostatic pressure to which the pipe is exposed during installation. The D/t values required will decrease to approximately 15-20 for 3,000 feet of water as compared to a range of 40-50 for 300 feet of water.

It is expected that presently available grades of pipe can be used in deep water application, since for low D/t's the nominal yield strength becomes less important than the ultimate strength of the pipe. For low D/t's (less than 40) the failure of the pipe under combined bending and pressure occurs in the plastic range, i.e., beyond the elastic limits. The nominal yield stress has little significance and strain controls the extent to which the pipe can be loaded.

From a manufacturing standpoint, the primary concern about the pipe expected to be required for deep water application is the heavier wall thickness, particularly in the larger pipe diameters. Almost all line pipe 30-inches in diameter and larger is welded pipe produced by the submerged arc process. Manufacturing is limited by the capacity of the plate forging equipment, which can produce heavier wall thickness pipe in low strength steel than in high strength steel, such as X-70 and above. Industry surveys show that 30-inch pipe suitable for laying in 3,000 foot depths will be available in 1980.

Pipeline Coatings

Most submarine pipelines are coated in order to provide corrosion protection and stability in the presence of sea bottom currents. In deep water, the heavy wall thickness required, together with concrete coating, can lead to increased pipeline submerged weight. In some cases, this could mean the line can not be laid by conventional methods. If the weight coating is omitted, the ability of the corrosion coating alone to withstand and transmit the higher tensions required when laying in deep water becomes an important question. Apparently, thin film epoxy coatings offer advantages in this respect over conventional mastic and coal tar enamel coatings.

Variations in pipeline submerged weight in deep water is an important consideration during installation because the unsupported span lengths may reach 3,000 feet instead 300 feet. The tolerances in coating density and thickness will need to be more closely controlled.

Improved Analytical Capabilities

Among the most important advances made in recent years are a better understanding of pipe properties and behavior, a better definition of the various loading phenomena the pipe undergoes during installation, and the development of improved analysis techniques and computer programs to aid in pipeline design and construction engineering.

Non-Linear Pipe Stress Analysis

Deep water will require that advantage be taken of the reserve strength of the pipe, which calls for an understanding of the pipe's non-linear or inelastic behavior. When plastic deformation occurs in a pipeline being laid, a residual curvature remains which could result in an undesirable final configuration on bottom. The residual out-of-roundness reduces the pipe load carrying capability during installation. Residual curvature will vary depending among other things on pipe weight, stiffness and the history of axial force and bending loads applied during installation.

A detailed analysis of the pipeline accounting for non-linear effects is required to ensure that residual strains are held to acceptable values.

Analytic tools have been and are being developed to predict pipe behavior in the inelastic region. With the availability of these tools, it is possible to use strain criteria rather than stress criteria in pipe design. This is not only a more logical approach, but also tends to permit greater confidence in the use of shorter pipe laying stingers, lower tension, heavier submerged weights, and less restrictive lay barge mooring criteria.

Dynamic Pipeline Stress Analysis

Concern over higher strain levels, longer unsupported spans while laying and fatigue effects makes a dynamic analysis of pipeline installation a necessity for deep water. Consideration must be given

to sea state, vessel motions, pipe D/t and weight, stinger configuration and pipe tension in order to evaluate the interaction
of all these variables. The output from this analysis may be
combined with a buckling analysis to determine the factor of safety
for a given set of conditions. Dynamic analysis is also essential
for any meaningful fatigue analysis. Concentrated industry efforts in
the past few years have led to the development of several sophisticated computer codes for dynamic analysis of pipeline and stinger.

Pipeline Fatigue Analysis

Fatigue caused by variations or reversals in loading has not
been a significant problem in pipeline installation to date. However,
fatigue must be considered for deep water because of higher strain
levels and a longer pipe exposure time in the suspended pipe string.
Due to the high strain levels and relatively few cycles of loading,
any fatigue problem that occurs would more likely be low cycle rather
than high cycle. The availability of suitable dynamic analysis
programs, together with improved knowledge of vessel motions, sea
states and material behavior make possible the assessment of any
potential fatigue problems while laying.

Pipeline Buckling Analysis

In the last three to four years, considerable work has been done
on plastic buckling of pipe, particularly for the low D/t ratios to
be used in deep water. In particular, more knowledge has been gained
of the "propagating" buckle phenomenon. This occurs whenever a
transverse buckle is transformed to a longitudinal buckle and pro-
pagates along the length of the pipeline (Figure 9). If the pipeline
is designed to resist a propagating buckle, the wall thickness and
weight may make it more costly and difficult to install by conventional techniques. If this phenomenon is ignored, the result could
be the loss of many miles of pipeline. A design which provides
buckle arrestors at intervals along the pipeline is the best
compromise, and this approach is now common in today's pipeline
construction projects.

Vessel Motion and Mooring Analysis

In deep water, the ability of a conventional anchor mooring
system to provide the necessary control of the lay vessel becomes
less certain. The development, analysis and design of improved
mooring systems is a major task crucial to the feasibility of laying
pipe in deep water using the lay barge method. Systems under
consideration for deep water include both active and passive winch
systems; anchors plus thrusters; and complete dynamic positioning
without anchors. All of these systems will utilize to some degree

COURTESY OF BATTELLE COLUMBUS LABORATORIES

PIPE DAMAGED BY PROPAGATING BUCKLE

FIGURE 9

on-board computers for data analysis and display, critical computations and mooring system control. The Saipem vessel, Castoro VI, has a computer controlled system for winches and thrusters. Other contractors are known to be developing similar advanced systems.

The ability to accurately predict the moored response of a vessel is linked very closely to the accuracy with which wind, wave and current forces acting on the vessel can be compared. This area of vessel motion analysis has been dramatically improved in the last few years through the development of sophisticated computer programs.

Stinger Analysis

In order to lay pipe in deeper water it may be necessary to go to longer pipe laying stingers than are presently being used. Even if this length increase is minimal because of higher tension capacity or increased stinger curvature, the cost of stinger and subsequent pipe damage and repair can be prohibited. Therefore, an understanding of the dynamics of the stinger and its interaction with the pipe is essential. Fortunately, improved analytic techniques and extensive model testing have provided better understanding in this area. These developments, coupled with improved vessel motion analysis techniques, provided the industry with engineering tools which were not available during the development of much of the existing equipment.

Pipeline Construction Simulations

With the sophisticated computer programs that have been developed to take advantage of today's high-speed computers, it is now possible to simulate by computer all important aspects of a pipelay constructing job, such as welding, weather and mechanical downtime, anchor moving, and various randomly generated occurrences. This permits evaluation of various equipment spreads for a given job as well as assessing of the probability of successful completion within a given time frame and cost range for a given set of circumstances (Figure 10). This is a powerful tool not yet in widespread use, but expected to be integral to planning deep water projects.

Special Problems

It has been said that oil and gas have been found in all the easy places. The remaining areas for extensive exploration and exploitation are characterized by special problems which will be very expensive to solve. Included are areas subject to severe environmental conditions, ice, seismic activity and mud slides.

150

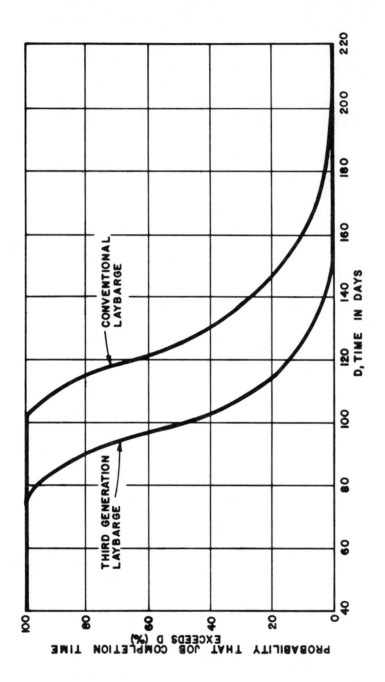

SIMULATED COMPARISON OF PIPELINE INSTALLATION TIME

FIGURE 10

Sites With Ice Coverage

The degree to which pipe laying activities will be subjected to surface ice depends upon the locale. In some areas, ice coverage is nil for a significant portion of the year, permitting near normal construction activities. In other areas, ice coverage varies throughout the year, calling for special construction techniques. Some areas are completely covered by ice during most years, which requires an entirely different technology for pipeline installation. (Figure 11). This includes working through the ice with equipment consisting of a cross breed between cross-country and offshore pipeline equipment. Ice breaking pipelay equipment, surface effect vehicles, bottom pulling techniques and modified conventional offshore laying techniques have all been studied for work in ice areas.

Site With Seismic and Mud Slide Activity

There have been nationally published photographs of the Alyeska Pipeline which show the geometric offsets, special supports, special instrumentation and controls and other techniques utilized to minimize the possibility of environmental damage in case of damage to the pipeline caused by seismic activity (Figure 12). It is highly unlikely that most of these measures can be economically adapted for submarine pipelines. Nevertheless, assurance will be demanded that submarine pipelines have been designed to withstand earthquakes to the extent that environmental damage will not be a major problem. Where practical, rerouting of the pipeline to avoid known fault zones will minimize risks, but added length of pipe must be considered in any economic analysis. Special "weak link" joints with check valves have been installed in some areas where bottom damage probability is higher than normal.

It is commonly said that if a large segment of the sea bottom wants to move due to a mud slide, a 12-inch pipe will not prevent it. Rerouting of the pipeline to avoid mud slide prone areas is the best solution presently available, together with the "weak link" joints previously mentioned.

Conclusion

Although the early years of offshore pipelining consisted primarily of an extension of land pipelining techniques to floating barges, the technological advances that permit us to work in hundreds or thousands of feet of water can hardly be considered mere extrapolations. While many countries--Norway, Italy, France, U.K., Holland, Japan--have made contributions to offshore technology in recent years, the foundation for the industry was laid in the United States.

COURTESY OF POLAR GAS

PIPELAYING IN ICE COVERED AREAS
FIGURE II

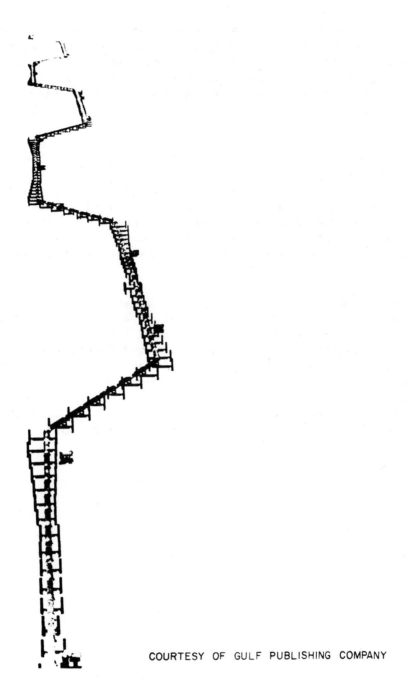

COURTESY OF GULF PUBLISHING COMPANY

PIPELINE DESIGNED TO WITHSTAND
SEISMIC ACTIVITY

FIGURE 12

The American free enterprise system has generated the risk capital required to finance the research, development and experimentation that has opened the ocean frontiers. American companies compete with one another, with foreign government subsidized companies, and sadly, sometimes with our own government and vocal special interest groups to provide the oil and gas products that are the life blood of our economy.

Exploration and production have been on-going off the Louisiana and Texas coasts for over 30 years without any provable adverse permanent effect on our environment. Commercial deep sea fishermen head for offshore rigs and platforms when they ply their trade. As long as man and machinery are involved, there will continue to be occasional accidents. Are the benefits worth the risk? Are the prices we pay for petroleum products unreasonable in terms of risk, return on investment and return on sales?

Reasonable people can honestly hold divergent views on these issues, but it is submitted that unless we are prepared to bring the economy and improvement of the standard of living of the western world to a grinding halt, we must lessen our dependence on foreign sources of oil and gas. The evidence is overwhelming that American companies are continuing, and will continue to invest in, the required programs of technology advancements to insure we have the knowledge to tap our ocean resources. The question is whether we have the national commitment to permit them to do so.

Mr. Lochridge, Vice President of Brown & Root, Incorporated, graduated from Rice University with BA and BSCE degrees. He holds sixteen United States patents and foreign counterparts. His present assignment is management of two engineering departments engaged in research and development and engineering studies related to offshore construction. He is also responsible for the Microwave Survey Group engaged in worldwide marine survey and communications.

OFFSHORE STORAGE, TANKER LOADING, FLOATING FACILITIES

D. M. Coleman

As tankers became larger after the closing of the Suez Canal in 1967, new types of offshore loading terminals were developed. Many navigation channels and harbors were inadequate for the drafts of the VLCC's (Very Large Crude Carriers) so offshore loading terminals were developed for use at exposed ocean locations.

Offshore storage was first needed in the Middle East where favorable sea conditions exist. Subsequently, in the 1970's gravity structures were being developed for the North Sea to provide drilling and production platforms. They were also designed to provide storage which permitted continuous production when weather conditions are too severe to conduct tanker loading operations.

Subsea pipelines normally transport crude oil from production facilities offshore, to onshore storage terminals and then on to markets. However, the severe North Sea environment and the deep 1,000 foot Norwegian trench presented serious technological challenges for timely installation of subsea pipelines. Consequently, alternative tanker loading facilities capable of operating in severe ocean environments were developed.

With the increased demand and sharp rise in prices for crude oil since 1973, there have been incentives to produce offshore marginal fields and to initiate production from a field as early as possible. New concepts in floating production platforms are being developed to accomplish these goals. In particular, conversion of semisubmersible drilling vessels and conversion of crude oil tankers have provided floating platforms for necessary production, storage and/or offload facilities. These have been used with single point moorings (SPMs) and shuttle tankers to provide a complete production system.

Selected examples are presented to indicate how offshore storage of crude and offshore storage with tanker offloading has been accommodated. Also, examples are presented to indicate how the complete system with storage and offloading to include the production facilities have been developed to serve the needs of industry.

OFFSHORE STORAGE

There are two basic types of offshore storage: those fixed to the seabed, and floating storage systems. Examples of each type are to be discussed.

Fixed Type Storage

Some of the earliest developments in offshore crude oil storage were the bottom-supported or fixed type. Figure 1 shows examples: Khazzan tanks, Ekofisk tanks, and Condeep structures.

Khazzan Tanks

The first large volume underwater crude oil storage system was installed in the Persian Gulf in 155 feet of water in 1969, by DUBAI Petroleum Company, a subsidiary of Conoco. This unique facility is entitled Khazzan I, a 500,000 barrel underwater storage tank, designed and fabricated by Chicago Bridge and Iron. The tank is a steel structure, 270 feet in diameter with an overall height of 205 feet. The lower section of the tank consists of a roof having an arc length radius of 180 feet that intersects with a cone transition to a column 96 feet tall. The tank is fixed to the sea floor by thirty 90 foot piles drilled and grouted in place. The Khazzan I was the first of three of these tanks to be placed in the Fateh field. Khazzan II and Khazzan III have added decks which support production facilities of 300,000 bbls of oil daily.

Ekofisk Tank

Ekofisk was the first commercial oil field discovered in the North Sea. Its four exploratory wells were subsea completions that produced to a modified jack-up drilling platform.

The field is located in the Norwegian sector of the North Sea and is separated from Norway by a 1,000 foot deep trench. Industry has not yet demonstrated the capability to maintain large-diameter lines across this trench; therefore, oil produced from the Norwegian sector must now either be piped to the United Kingdom or offloaded to shuttle tankers.

Early production at Ekofisk was subject to weather shutdowns because the subsea/jackup arrangement lacked storage facilities. Development of the field continued from fixed steel platforms. To provide storage, Phillips Petroleum, the operator, installed a million-barrel oil storage facility which was designed by C. G. Doris, and fabricated by two Norwegian firms, Selmer and Hoyer-Ellefsen. The 302 foot diameter prestressed concrete oil storage tank is a nine module storage unit, surrounded by a concrete breakwater of the Jarlan perforated caisson type. It was constructed in a Norwegian fjord, towed to its Ekofisk location, and installed in 230 feet of water. It became operational in the summer of 1972. Later in the field development its two decks served as the base for a 300,000 barrel/ day oil and gas processing center, from which oil is transported by pipeline to the Teeside Terminal in Scotland.

Condeep Structure

The North Sea has soil conditions consisting of very stiff clays and dense sands which are able to support heavy gravity platforms. The first Condeep was installed in August 1975 in Mobil's Beryl Field in water 380 feet deep. The Beryl "A" Condeep base consists of 19 vertical reinforced-concrete cylinders, each 66 feet in diameter. Fabrication was in Norway and the platform was towed to the field with most of the facilities on the deck.

Sixteen of the cells are 154 feet high and provide 900,000 barrels of oil storage. The three remaining cells extend above the ocean, forming three legs which support the steel structure for drilling and producing operations. Tanker loading facilities at Beryl "A" are provided for by an ALP (Articulated Loading Platform), which transfers the stabilized crude oil to 80,000 dead weight ton tankers. Several other fields in the North Sea also have Condeep gravity structures with offshore crude oil storage capacity to a million barrels.

Floating Type Storage

Floating type storage units shown on the bottom half of Figure 1 have been used at many locations because existing barges and tankers were readily available for conversion to storage. As industry moves into more severe environments, new shipyard construction may be required for redesign using the established methods of ship construction.

158

OFFSHORE STORAGE

FIXED TYPES

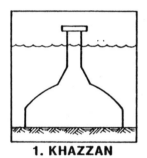

1. KHAZZAN

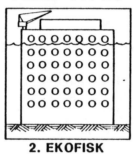

2. EKOFISK

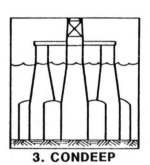

3. CONDEEP

FLOATING TYPES

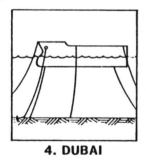

4. DUBAI

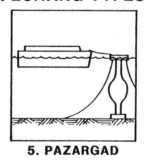

5. PAZARGAD

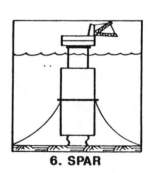

6. SPAR

FIGURE 1

DUBAI Floating Storage

In the Fateh Field located in the Persian Gulf in water depths in 190 feet, DUBAI Petroleum used two tanker forebodies for floating storage as a temporary facility. Two small disabled tankers were purchased and converted to storage facilities providing about 250,000 barrels of storage. Spread mooring was used for the two facilities.

Pazargad

The Iran Pan American Oil Company used a barge facility to provide crude oil storage for its Cyrus field also located in the Persian Gulf. Initially, a small 40,000 DWT tanker was placed in temporary service to provide crude oil storage. In 1970, this tanker was replaced with Pazargad, a 750,000 barrel barge illustrated on Figure 1. Pazargad was designed and built by Mobil and at the time was the world's largest crude oil barge. Pazargad is moored by a single buoy mooring system in 140 feet of water. Crude is transferred from storage through a submarine line to an export tanker via an SPM about a mile away.

SPAR

In 1976, the Shell-Esso Brent field in the North Sea used the SPAR as developed by Shell and IHC Holland to provide crude oil storage. SPAR shown on Figure 1, is a vertical floating storage tank 450 feet high with a 96 foot diameter. It provides 300,000 barrels of crude oil storage and is stationed in 460 feet of water with a six leg catenary mooring. The SPAR type of facility is attractive because it is reasonably stable in high seas. Another unique feature is its loading capability for medium sized (70,000 DWT) tankers, as will be discussed later.

Tanker Loading

Single Point Moorings

For tanker loading at sea, the Single Point Moorings (SPMs) were developed for offshore locations where the wind, waves and currents have directional variability. This type of mooring allows the tanker to weathervane, thereby reducing the mooring forces. A number of different types of single point moorings, as shown in Figure 2, are in use throughout the world. These include the Catenary Anchor Leg Mooring (CALM), Exposed Location Single Buoy Mooring (ELSBM), Single Anchor Leg Moorings (SALMs), Articulated Loading Platform (ALP), and again the SPAR which also provides storage.

160

TANKER LOADING - SINGLE POINT MOORINGS (SPM)

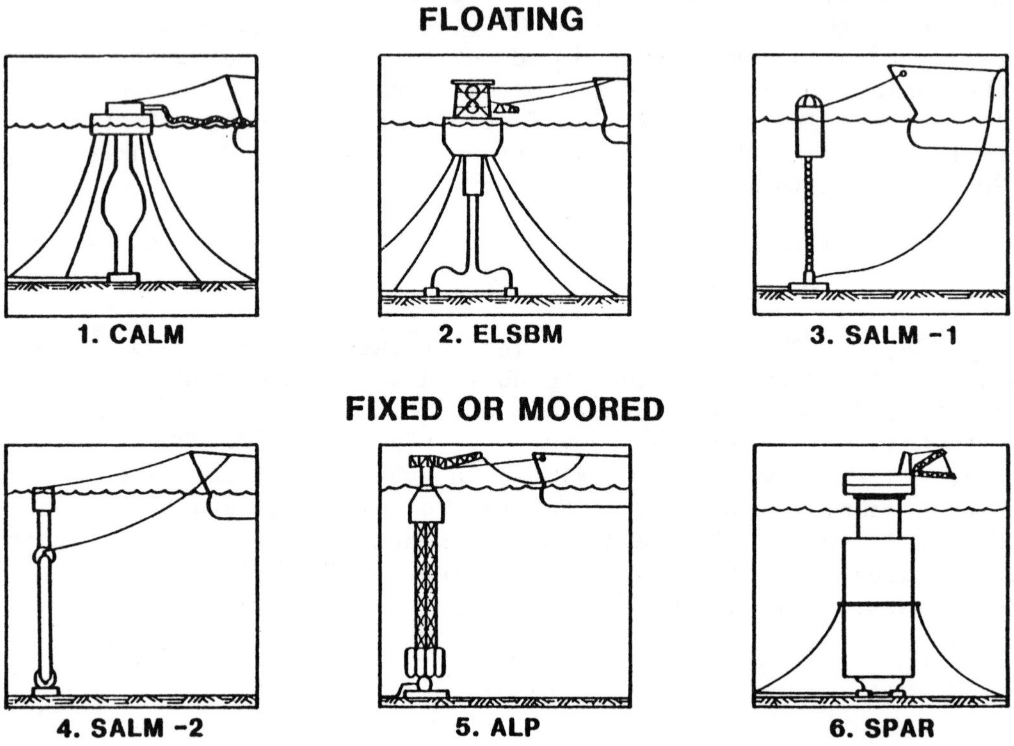

FLOATING

1. CALM 2. ELSBM 3. SALM -1

FIXED OR MOORED

4. SALM -2 5. ALP 6. SPAR

FIGURE 2

The first SPM scheduled for use in United States waters will be Exxon's Hondo field SALM off California. The second SPM application will be the SALM used by the Louisina Offshore Oil Port (LOOP).

Catenary Anchor Leg Mooring (CALM)

The first type of SPM developed was the CALM by IMODCO and Shell. The CALM terminal is used at locations with shallow to moderate water depths and placid to moderate environmental conditions. The CALM (Figure 2) is basically a disk shaped surface buoy 30 to 45 feet in diameter and 10 to 17 feet high with spread mooring. The spread mooring is made up of multiple chain anchor legs and is anchored to the sea-bed by conventional marine anchors, clump weights, or piles. Crude oil is transferred by submarine pipeline connected to the buoy with flexible underwater hoses. A fluid swivel on the buoy is connected to floating hoses for loading tankers.

During the summer months of 1979, a CALM was installed in 369 feet of water at British Petroleum's Buchan Field in the North Sea. This CALM was designed by IMODCO, and is the deepest and most severe environment application of the CALM thus far in the North Sea.

ELSBM

A second generation CALM is the ELSBM shown on Figure 2. The unit is more stable due to its surface buoy configuration, which is a modified spar-disk buoy type. Shell has used this design at its AUK field in the North Sea.

Single Anchor Leg Mooring (SALM)

The SALM was developed in the mid-1960's by Exxon, when it became apparent that new concepts would be needed to moor and load the larger tankers. The SALM-1 shown in Figure 2 consists of a mooring buoy at the surface and is attached to a base on the seabed by a single anchor chain or leg. The buoy is drawn down against its buoyancy and provides tension in the anchor leg. Oil is pumped through a flexible hose from the sea bed onto the tanker. The first SALM was installed by ESSO Libya at Brega in 1969 in 140 feet of water, where it could moor very large crude carriers (300,000 DWT).

SALM -2

The world's deepest SALM was designed by Single Buoy Mooring, Inc. and installed in 1977 in 530 feet of water in British National Oil Company's Thistle Field in the North Sea. It has a gravity base connected by an articulation to a 335 foot riser, which in turn is connected by articulation to a cylindrical surface buoy. Flexible hoses bypass the articulation on the base. A fluid swivel is located in the bottom of the surface buoy, to which the loading hoses are connected by rigid piping. The Thistle SALM can moor large (120,000 DWT) tankers.

Articulated Loading Platform (ALP)

For the severe North Sea environment, Mobil chose the ALP designed by the French firm EMH. The Beryl location is in 385 feet of water and can moor medium sized (80,000 DWT) vessels. The Statfjord location is in 475 feet of water. The ALP also shown on Figure 2 consists of a base structure and a universal joint with two axis' of rotation connected to the vertical column. The oil flows through the universal joints at the base. At the top of the column a structure with rotating head and a 120 foot flow boom to support a 16 inch loading hose is provided. On occasion, loading operations have continued without interruption in 50 to 60 knot winds and waves to 30 feet high.

SPAR

Note the SPAR in Figure 2 is again shown as a moored type of SPM, but since it also provides storage it will be discussed later.

Combined Storage and Loading Facilities

Most of the combined storage and loading facilities consist of a tanker or barge moored to a SPM by a rigid yoke as shown in Figure 3. Shuttle tankers are loaded alongside or in tandem. Other combination facilities are single self-contained units, and numerous existing installations and proposed new designs fall in this category.

Single Buoy Storage (SBS)

The Single Buoy Storage (SBS) consists of a modified CALM, a rigid mooring yoke and storage tanker as shown in Figure 3. The yoke-moored tanker has numerous characteristics which make it more attractive than a hawser-moored tanker. Small to large storage tankers are presently in use.

One of the earliest SBS installations was completed in 1974 in the Ashtart field, Tunisia for Aquitaine Tunesie. The SBS provides crude oil storage of about 400,000 barrels on a medium sized tanker. Smaller shuttle tankers are loaded when moored along side.

COMBINED STORAGE & TANKER LOADING

EXISTING

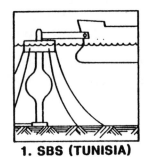

1. SBS (TUNISIA)

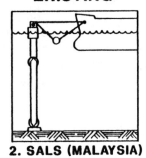

2. SALS (MALAYSIA)

3. SPAR (BRENT)

NEW DESIGNS

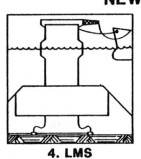

4. LMS

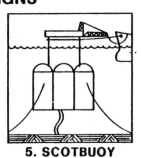

5. SCOTBUOY

FIGURE 3

Single Anchor Leg Storage (SALS)

The mooring of a storage tanker by rigid yoke to a SALM (Figure 3) is referred to as a Single Anchor Leg Storage (SALS). In 1978, the first SALS using a VLCC was installed in 214 feet of water in the Pulai Field, offshore Malaysia for Esso Production. The tanker has a storage capacity of one million barrels. Loading of small to large shuttle tankers is accomplished by bringing them alongside and berthing to the storage tanker.

SPAR

The SPAR referred to previously is a system which combines storage and offloading capabilities. The offloading function is provided for by a retractable loading boom mounted on a turntable. The tanker is moored by a bow hawser to SPAR, and loading operations can accommodate small to medium sized tankers. The Shell-Esso Brent field in the North Sea used a 300,000 barrel SPAR for tanker loading prior to completion of the pipeline to the United Kingdom.

New Designs

Many new concepts are presently being developed for combining storage and offloading. Two, which have been carried through preliminary engineering design stage, will be discussed.

Loading/Mooring/Storage Vessel

The Loading/Mooring/Storage Vessel (LMS) shown in Figure 3 developed by Mobil, is a unique concept, which provides storage capacity of 1-1/2 million barrels and can load VLCC tankers. The LMS is all steel construction, and uses an innovative, diagonally suspended, synthetic rubber diaphragm in each tank to isolate crude oil from clean water ballast. The motions of the LMS are such that loading of tankers can be conducted during severe environmental conditions in the more hostile areas, thus resulting in high lift efficiencies.

SCOTBUOY

SCOTBUOY also shown in Figure 3 was developed by Seven Seas Engineering from Scotland and is a storage/loading system which is composed of vertical concrete cylinders moored to the seabed. A steel upper section supports deck and loading facilities. This concept has been proposed also to provide service in hostile environments.

Floating Production Facilities
Semi-Submersible Production Facilities

Argyll Field

The first Floating Production Facility (FPF) as shown in
Figure 4 commenced oil production in 1975, in the North Sea's Argyll
Field. The Argyll field was developed by Hamilton Brothers, using
subsea wells with a tensioned production riser connected to a
converted semi-submersible drilling vessel moored in 250 feet of
water. Production equipment was designed for a capacity of 70,000
barrels of oil per day (BOPD) and is located on the decks of the
semi-submersible. The field is currently producing 36,000 BOPD.
Produced oil is sent from the FPF via the production riser to a CALM
terminal, where small (55,000 DWT) tankers are loaded.

Enchova

The Enchova field, offshore Brazil, is currently being produced
with two FPF's: the SEDCO semi-submersible shown in Figure 4 handles
a production rate of 10,000 BOPD which is produced to a very small
(12,000 DWT) storage tanker. This spread-moored storage tanker then
offloads alongside to shuttle tankers. The Penrod semi-submersible
handles a production rate of only 5,000 BOPD which goes to a CALM and
a small tanker.

Ship-Shape Production Facilities

Castellon Field

In 1977, Shell Espana put into operation an innovative Floating
Production Facility at the Castellon field located in 380 feet of
water about 40 miles offshore Spain in the Mediterranean Sea. A
subsea well is produced to the FPF with a rigid buoyant yoke attached
to a small (60,000 DWT) tanker. The tanker provides storage and
facilities to produce about 20,000 BOPD.

Garoupa Field

An ambitious early production system was considered by Petrobras
in 1974 for the Garoupa field offshore Brazil. This early production
system was to include nine subsea wells at water depths from 300 to
500 feet, completed with Lockheed atmospheric well-head chambers
connected to an atmospheric manifold chamber on the seabed. Crude
oil is piped from the wells to the manifold center through an Anchor
Leg Processing tower (ALP). This articulated tower is shown in
Figure 4 and provides for the mooring of a small dedicated tanker
with processing facilities and gas flaring capability.

FLOATING PRODUCTION FACILITIES

SEMI-SUBMERSIBLE INSTALLATIONS

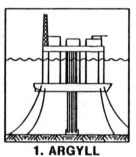
1. ARGYLL

2. ENCHOVA

SHIP-SHAPE INSTALLATIONS

3. CASTELLON

4. GAROUPA

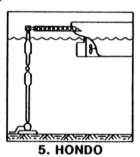
5. HONDO

FIGURE 4

Processed crude is shipped backed down the riser to another ALP for tanker loading. Both towers were designed and fabricated by Chicago Bridge and Iron.

Hondo Field

The development of the Hondo field in the Santa Barbara Channel by Exxon will be the first Floating Production Facility used in the United States. The Hondo field has a 28-well fixed steel platform set in 850 feet of water, and a SALM (Figure 4) set in 490 feet of water, located 1-1/2 miles from the platform. The SALM, used to stern-moor the 50,000 DWT tanker will be the first SPM to be used in the United States. It was designed and constructed by IMODCO of California. Gas-oil separation is accomplished on the nearby fixed platform, where the gas is dehydrated and compressed for injection. The process facilities on the tanker will handle the water-oil separation, and initially will have a capacity of 55,000 BOPD.

Exxon has been trying to get the Hondo field on production for over 10 years, but has been delayed by the various regulatory bodies. It has now resumed construction of the various components of the production and transfer system. The offshore storage and treating vessel is being outfitted in Texas, and the SALM is being fabricated in California. Exxon recently gave these reasons for the delay: three major environmental impact studies, 21 major public hearings, 10 major government approvals, 51 studies by consultants and 12 lawsuits. Hopefully, the Hondo installation will start up in 1981.

Combined Production/Storage/Loading Facilities

Combined facilities incorporating production, storage and tanker loading capabilities have become attractive for a number of reasons. For example:

- One floating platform can accomplish all three functions.

- Since lead time for design, construction and installation is shorter than alternative methods, earlier production of a field is possible.

- Storage and loading facilities are provided for fields without pipelines.

- The economic production of marginal fields is improved.

COMBINED PRODUCTION, STORAGE, TANKER LOADING

EXISTING

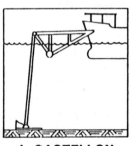

1. CASTELLON

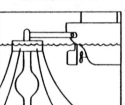

2. NIDO

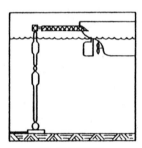

3. HONDO

DESIGNS

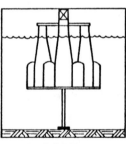

4. CONPROD

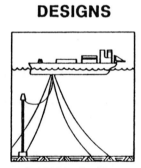

5. SHIP-SHAPE

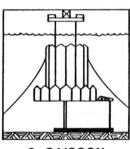

6. CAISSON

FIGURE 5

Three cases are presented, describing different locations where this new concept has been utilized.

Castellon Field

The Castellon field installation offshore Spain (Figure 5) as described earlier is able to incorporate production facilities on a small storage tanker; because only small quantities of gas are produced which are disposed of with incinerators. Loading operations are conducted by mooring small shuttle tankers alongside and offloading the crude oil from storage.

Nido Field

Cities Service uses a medium sized (100,000 DWT) tanker to provide production, storage, and offloading facilities in the Nido field in the South China Sea. As shown in Figure 5, the tanker is moored by the stern to an SBM yoke and mooring buoy, and has processing capacities of up to 50,000 BOPD. Gas is flared from a fixed platform nearby, with flash fuels burned in the tanker's boilers.

Loading is provided by a shuttle tanker mooring to the bow of the storage tanker. This is the industry's first end-to-end loading operation.

Hondo Field

The Hondo field installation, as described previously, will include a tanker FPF moored by a SALM (Figure 5). The 50,000 DWT tanker, in addition to its process facilities, will provide crude oil storage and loading facilities. At present, tandem loading of shuttle tankers is being proposed.

New Designs

Many new concepts are presently being designed for floating production/storage/loading facilities that are required for environmentally hostile areas of the world, such as the northern North Sea, and the far North Atlantic. Also for deep water locations, such as offshore California, the 6,000 foot deep East Coast Reef and the Exmouth Plateau off Australia. Several of the concepts which have been carried through preliminary engineering design stage are to be discussed.

Conprod

Conprod (illustration 4 in Figure 5) is an all-concrete floating platform, designed by Norwegian Contractors, which provides for oil production and storage at deepwater locations. It is based on the same technology as the Condeep structures. Conprod consists of submerged concrete cells, 82 feet in diameter and 177 feet high, for crude oil storage; and tapered columns, 344 feet high, to support the deck structure. A tensioned production riser is proposed to be used in conjunction with subsea well completions. Direct offloading to tanker is envisaged.

Ship Shape Production Platform

Mobil is developing a deep water production system (illustration 5 in Figure 5) with a very large crude carrier (1-1/2 million barrels of storage) to provide production/storage/loading facilities for large oil fields located in moderate to severe environments and water depths to 2,500 feet. A unique turret mooring system is being designed to be located forward of amidships. The turret mooring system allows the vessel to weathervane in response to the directional variability of wind, waves and currents, and also provides a non-rotating moonpool for riser handling. A compliant production riser in conjunction with a Subsea Atmospheric Well System will be used to produce the field. Loading of large shuttle tankers is to be accomplished by the tandem loading method.

Floating Caisson Vessel

Exxon is currently developing the caisson vessel system (illustration 6 in Figure 5) for use in water depths of 1,900 to 4,900 feet. It is designed to support drilling and processing facilities for 175,000 BOPD with 600,000 barrels of crude oil storage for use with medium sized shuttle tankers. Loading will be from a rail mounted carriage loading facility. A production riser for sub-sea wells would be connected to a Submerged Production System.

Conclusions

The following conclusions are presented for your consideration:

1. Industry has responded to the need to produce and transport oil from fields located in ever increasing water depths and more severe environments.

2. Although not the subject of this paper, it should be obvious from the description of the examples cited that costs for the facilities will increase with water depths and that risks will be greater to operate in the more severe and hostile environments.

3. Most of the offshore storage, tanker loading and floating production facility technology has been developed for fields located in foreign waters with fabrication and construction in foreign yards with very few installations in American waters.

4. Industry has responded by developing the offshore technology required for the 1970's and is well prepared to meet the increased requirements of frontier technology for the 1980's.

Mr. Coleman is Offshore Engineering Manager, Mobil Research and Development Corporation, Dallas, Texas, a position he has held since 1975. He has been with Mobil for over 30 years. He holds a BS in petroleum engineering from the Colorado School of Mines (1949) and an MS in petroleum engineering from Texas A & M (1954).

SUPPORTING SYSTEMS FOR INSTALLATION, INSPECTION, AND REPAIR

Michael D. Hughes

Ladies and gentlemen, I am in the business of performing work under water. My industry, which only a few years ago was limited to work in less than 300 feet of water, can now provide safe and cost effective services for any depth presently being considered for outer continental shelf development.

The purpose of my presentation is to acquaint you with the state of the art in underwater work and the expected capabilities of the near future. As you have seen here today, great progress has been made in developing safe and economical methods and equipment for OCS exploration and production. Many of these methods allow drilling and production to proceed without a requirement for manned intervention under water.

Breathing ordinary compressed air, we could work safely and efficiently to about 200 feet in the past. About 15 years ago, the development of saturation diving techniques by the U.S. Navy and further development by private industry extended our effective work capabilities to about 700 feet. Today we have clearly proven our ability to dive safely and effectively to at least 1,600 feet. Work at 2,000 feet is probable, and divers have performed effective work at a simulated depth of 1,600 feet in a laboratory. Therefore, the capability to work in the open ocean at this depth certainly exists.

At these greater depths, however, the risk to human life and the cost of task performance increase significantly. I will give you an example: One manned dive 850 feet deep from the Hondo structure off California would cost Exxon about $200,000 for five minutes of underwater work. This has led to the development of a whole family of diver alternative work systems (DAWS) that do not present as much human risk and extend our work capabilities far beyond the range of conventional diving, while reducing the financial costs. I would like to familiarize you with these methods and their capabilities.

There are two major categories of underwater work systems. Small manned submersibles have been used for a wide variety of tasks, and several are currently in service assisting OCS exploration and production. I believe most of you are relatively familiar with these small submarines and their capabilities.

One interesting fact you may not know is that these submersibles are currently capable of working at depths as great as 8,300 feet. They have actually performed drilling rig support work as deep as 4,876 feet.

The other major category of diver alternative work systems is remotely controlled vehicles. These vehicles range in size from that of a basketball up to vehicles large enough to carry three manipulator arms and lift several tons from the sea floor. Operated from the surface, these vehicles totally eliminate risks to human life, while providing many of the capabilities offered by manned intervention.

The principal problem confronting the engineers and managers conducting OCS frontier exploration and production is not whether the capability exists to intervene under water if necessary to perform inspection or repair operations, but rather the problem of choosing the safest and most efficient system from the variety available. Figure 1 presents an example of the considerations in selecting a system.

Figure 2 ranks various diver alternative work systems relative to functional utility, safety and cost. It shows some of the considerations across the top. I might add that Exxon was very happy to hear that we can perform the same task on its Hondo platform with an atmospheric diving suit at a cost of $20,000 or one-tenth, of a manned dive.

I would like to close with a few important statistics to prove that these methods and systems are not just concepts or unproven prototypes. We currently employ 17 Jim suits and WASPS. Some of these have been in service for over five years. The suits presently in the field have depth capabilities up to 2,000 feet. The current joint design is adequate to 3,000 feet. Production of 3,000 foot rated suits requires no new development, simply a requirement.

We have logged thousands of working dives at depths up to 1,440 feet. There are currently seven manipulator bells such as the arms bell in service with depth ratings as great as 4,500 feet. Several hundred working dives have been conducted at depths up to 3,300 feet. The first application for this bell was in 2,750 feet of water. The client originally estimated an average of approximately 1 1/2 dives per month. When the drilling operations began, they found that the bell was extremely useful and that it was quite easy and safe to operate. We have averaged 18 dives per month, and many of these at

TYPE OF DAWS

Work Conditions	Work Conditions	Untethered Sub. (Pisces)	Tethered Sub. ("Arms" Bell)	Atmos. Diving Suit-Thrusters (WASP)	A.D.S. (JIM) (From Platform)	RCS-Observer (RCV-225)	RCS-LTD Work (RCV-150)	RCS-Full Work (SAAB SUB.)
Work Conditions								
Maximum Depth	250 M	5	5	5	5	5	5	5
Maximum Currents	1 Kn	2	4	2	4	3	4	4
Minimum Visibility	2 M	3	4	4	4	4	3	4
Min. Access Space	30 X 30 CM	2	4	4	4	*	3	4
Max. Work Radius	.5 M	3	4	3	3	*	3	4
Typical Tasks	Task Difficulty							
Inspect/Observe	E	3	3	4	4	5	3	4
Recover Tools	E	2	4	3	4	*	3	4
Clean, Brush, Chip	E	2	4	3	4	*	3	4
Cut Cables	D	2	4	3	4	*	3	4
Jack, Spread	D	1	4	2	4	*	1	4
Untangle Lines	D	1	4	3	4	*	2	4
Attach Lines	D	3	4	4	4	*	3	4
Connect Hydro. Lines	C	1	4	3	4	*	1	4
Opr. Overrides	C	2	4	3	4	*	2	4
Open Close Valves	C	1	4	4	4	*	1	4
Stab Overshots	B	1	4	3	4	*	1	4
Make up Kill Line	B	*	4	3	4	*	*	4
Bolt, Unbolt	B	*	4	3	4	*	*	4
Replace Valves	A	*	3	3	4	*	*	4
Drill, Tap	A	1	4	3	4	*	*.	4
Place Shaped Charge	A	1	3	3	3	*	*	4
Precise Alignment	A	.1	4	3	4	*	*	4
Non-Destr. Testing	A	1	4	3	4	*	*	4
U/W Welding	A	*	4	2	3	*	*	4
Replace Modules	A	1	4	3	4	*	*	4
Precise Measurement	A	*	4	3	4	*	*	4
Midwater Observation	E	4	4	5	*	4	3	3

Scoring: TASK DIFFICULTY: A = Most Difficult, E = Least Difficult
System Capability: 5 = Most Capable, 1 = Least Capable
NOTE: * Incapability inherent in Design.

FIGURE 1: TENTATIVE APPLICATIONS CHECK LIST

176

TYPE DAWS		FUNCTIONAL UTILITY								SAFETY	COSTS	
		Inspection	Simple Repairs	Repair of Modules In-SITU	Equipment Salvage	Debris Clearance	Measurement and Alignment	Structural Repair	Midwater Cap	Risk to Personnel	Day Rate Cost	Long Lease Cost Per Day
MANNED	Untethered Manned Sub.	4	2	1	2	3	2	1	5	4	3	3
	Tethered Manned Sub.	3	4	4	3	4	3	3	3	3	3	4
	Atmospheric Diving Suit W/Thrusters (Mid-water)	4	3	3	4	3	3	2	5	2	4	4
	Atmospheric Diving Suit (on platform)	4	4	4	4	4	4	3	1	2	4	4
REMOTE CONTROL	Observer RCS*	4	*	*	*	*	*	*	4	5	4	5
	Limited Work RCS	3	2	2	4	3	2	1	4	5	4	4
	Full Work RCS	3	4	4	4	4	4	3	4	5	3	3
DIVER	Hyperbaric Diving	5	5	5	5	5	5	5	3	1	1**	1**

NOTES: Ratings Key: 5 = Most Favorable, 1 = Least Favorable

* Inspection system only, relative score does not reflect true value.

** Applies to 200M and greater depths. Cost is relatively less in shallow water.

FIGURE 2: TYPE OF DIVER ALTERNATIVE WORK SYSTEM VS FUNCTION, SAFETY, COST

depths of almost 3,000 feet. I think that gives you some indication as to the simplicity and effectiveness of this type of operation.

There are literally hundreds of remote vehicles in service with depth ratings up to 3,000 feet, and these can also be increased in depth rating quite easily. The small manned submersibles in service number several dozen that are capable of depths up to 8,400 feet.

Conventional diving is limited at this time to a probable maximum depth of 2,000 feet. Atmospheric diving systems are rated at 2,000 feet, could easily be extended to 3,000 feet. The point, of course, is that there is a wide variety of capability available for deepwater support. The capability exists today to provide underwater inspection, maintenance, and repair services safely and efficiently in any water depth presently being considered for OCS frontier exploration.

I would like to echo the comments of Joe Lochridge. I would like to remind each of you that this country's greatness was built on our willingness to intelligently use our resources and technology. Many of you here are in positions to influence the future use of our OCS and its underwater resources and technology. I hope that each of you will have the wisdom and courage to ensure that these are developed so as to reduce our dependence on foreign interests. We are ready to do the job if you give us the opportunity. Thank you.

DR. RICKARD: Thank you very much, Mike.
There have been a lot of large numbers tossed around today. The $200,000 for five minutes' worth of work was a new one to me, and it is beguiling to think of a company having to pay that cost, but I think it also behooves us to remember that in the private enterprise system, a company that incurs those costs must recover them. Traditionally, costs are recovered through the price of the product sold to the public. So in a very real sense, Exxon is a conduit, and it is society that eventually pays those costs. It is important to remember this.

Mr. Hughes is Executive Vice President of Oceaneering International, Incorporated, and is responsible for sales, operations, research and development and marketing. His experience includes the design of diving equipment and systems. He designed a self-propelled ditching machine, underwater tools, environmental control systems to be used with saturation diving systems, "space helmet" type breathing equipment, and supervised the first diving bell system in connection with drilling operations in the Gulf of Mexico.

Mr. Hughes began his career as a professional diver in the offshore oil fields of the Gulf of Mexico. He holds a BS in Civil Engineering from the University of Tennessee and is a member of numerous professional societies and associations.

OIL SPILL CLEANUP AND CONTAINMENT

Leon J. Kazmierczak

History

A heightened awareness of the environmental threat posed by oil spills was brought about by the Torrey Canyon and Santa Barbara incidents. I will not review those events here because their histories are well-known and adequately documented.

One result of the increasing concern over spill accidents was the passage of legislation seeking to prevent, minimize or mitigate such occurrences. A list of federal legislation would include the Clean Water Act of 1970 and subsequent amendments in 1972, 1977, and 1978; the Ports and Tanker Safety Act of 1979; the Outer Continental Shelf Lands Act Amendments (OCSLAA); the Alaska Pipeline Act; and the Deepwater Port Act.

Regulations pursuant to the OCS Lands Act Amendments require that a spill contingency plan be written, implemented, reviewed, and approved by the U.S. Geological Survey (USGS) before exploratory drilling may proceed on OCS Leases.

Immediately after the Santa Barbara spill, a number of major oil spill cooperatives were organized. Oil spill cooperatives, for those of you who are not familiar with the term, are organizations made up generally of member oil companies who pool their resources and finances for the purpose of acquiring equipment jointly so that they can share costs and provide the level of capability that is needed to operate both in harbor areas and offshore.

Another advantage of cooperatives is that they can provide reasonably quick response because, generally they are located in areas of concern; that is, where the members are actually conducting their activities.

To give you some idea of the magnitude of this movement, there are now over 100 cooperatives in the United States. A number of them are small, but there are also a number of major ones, and I will just list those for you: Clean Bay, which operates in San Francisco Bay; Clean Sound in Puget Sound; Clean Seas and the Southern California Petroleum Contingency Organization, both in southern California; Clean Gulf, which operates in the Gulf of Mexico at various locations; Clean Harbors, which is newly formed, operating in the New York Harbor area; the Delaware Bay Co-op, one with which I am most familiar; Clean Atlantic, which was formed for the purpose of protecting the OCS work in the Atlantic, all the way from the south and mid-Atlantic to the north Atlantic, out of Davisville, Rhode Island; and Clean Caribbean.

179

Aside from the oil companies' capabilities, there are many spill contractors who are located in sensitive areas such as harbors and producing areas. Spill contractors are hired to help clean up spills whenever there are such problems. Last but not least, by any means, is the Coast Guard which operates the strike teams that have been formed under the National Contingency Plan. They are located on the East Coast, on the Gulf Coast, and also on the West Coast.

The cooperatives mentioned above are funded at levels exceeding a million dollars and sometimes approaching $4 or $5 million. That represents a rather sizable commitment.

I would like to touch on a few other items. I am associated with the Oil Spills Committee of the American Petroleum Institute (API). The committee was formed late in 1969 or early 1970, and since that time it has sponsored an impressive body of research. We have funded over 50 individual projects, involving the development of equipment for containment and removal and also for examining fate and effects of oil in the marine environment. The Environmental Protection Agency (EPA), the Coast Guard, and the Navy have also conducted and continue to conduct studies in these areas.

Aside from research and development, we have been involved in other areas of interest. One was the establishment of an oil spills training school at the Moody Campus of Texas A&M in Galveston, Texas. Sessions run for a week and are conducted every other week. Since it began in 1975, almost 2,000 students have attended. Also, every other year, in cooperation with EPA and the Coast Guard, API sponsors a meeting that is truly international, where recent developments are discussed, and technology and information are exchanged.

One additional point should be made. Oil companies, the Coast Guard, and others who have the responsibilities for cleaning up spills get quite a bit of criticism for not being able to contain and remove oil in heavy sea states. However, there are two major considerations in this regard. One is that it is unsafe to operate under certain conditions, and that is a very practical consideration. Another practical consideration is that it is very difficult to find oil in 8 and 10-foot seas because there is so much energy being generated by the wind and the waves that the oil gets churned up into the water column. It is, in effect, unrecoverable.

Equipment for Recovering Spilled Oil

Figures 1 through 7 are of a demonstration of a unique boom design. In the past, booms built to operate in unusual conditions had to be rather meaty, very strong, and consequently become very heavy and hard to handle.

Now, this kind of equipment can be built with durability and yet it combines strength with lightweight characteristics. In Figure 1 the box contains 400 feet of boom. It is a compactable boom and compresses much like an accordion. The boom is a Seacurtain manufactured in Torrance, California.

Figure 2 shows an abrasion pad. It is poor practice to drag a boom across rocks or concrete or some other abrasive surface which can destroy the integrity of the boom. Therefore, it is helpful to use some kind of a plastic pad or a roller arrangement where the boom may be deployed across a roller which eliminates the scraping action.

The boom is beginning to emerge in Figure 3. This boom is open to the atmosphere. It self-inflates. It contains holes which are on top of the boom. When it is placed under tension and is pulled out, it sucks air in and self-inflates.

The structural rigidity is maintained by a stainless steel coil that is inside the boom. It can be seen in Figures 4 and 5 where the coil is under the folds of the boom covering. In Figure 5 the section to the left is fully inflated and the section on the right has not yet been stretched to that point.

Figures 6 and 7 show that this boom has good wave conformance charactistics; that is, it follows the water surface. This property is important for oil containment because it reduces splashover.

Figure 8 is a picture of a larger boom of the same design. This larger version was purchased by Clean Atlantic Cooperative for use in the north Atlantic. It has a 24-inch diameter float and a 3-foot skirt. It is heavy duty boom and, yet, light enough to be relatively easy to handle. Figure 9 is another picture of the same boom.

Figure 10 is a picture of a portable pump system. The Coast Guard has a number of these, or similar pumps. Others who own this type of equipment are some of the spill contractors. One will be purchased or leased by the Delaware Bay Co-op. I would imagine some of the other major coops also have this kind of equipment.

Figure 11 is another view of the system. It consists of a diesel engine which drives a hydraulic power train, which in turn powers a pump that has been lowered into the hull of a stricken vessel. If a vessel is stranded or grounded and it has no power, it can still be partially or wholly unloaded by using a pump like this. As the vessel is unloaded, it will float higher and in many cases free itself. Another plus is that oil which is recovered is no longer in danger of leaking.

Figure 12 is one of associated equipment where it is made possible for a vessel to come alongside another vessel through the use of very large and durable fenders which prevent hull damage.

Figure 13 is another type of inflatable boom. This is the Goodyear boom. It has individual inflatable compartments. It is durable and has good abrasion resistance. The individual inflation compartments make it difficult to recover and store for future use. Another problem with this type of boom is that it requires a source of compressed air, presumably either pressurized tanks or compressors.

The next series of figures document a demonstration that was conducted in the Delaware Bay on September 12 with the DIP 5001 skimmer vessel.

Figure 14 is a drawing of internal workings of this vessel. It uses an endless belt that rotates clockwise, using the hydrodynamic forces of the approaching water surface and oil film. The vessel could either move to the right or stay downstream of the current in a stationary mode where the wind or current will force the oil against the surface of the belt. It coalesces and then floats behind the belt mechanism into a collection well. There is an opening aft where water is released behind the vessel. The operation is depicted starting with Figure 15.

The DIP 5001 is a vessel fitted with folding doors that are closed for running mode, and open for an operating mode. These doors are similar to those on the landing vessels in the Second World War. They open on either side. There are sweeps that come down out of the hull, which are used to capture oil.

At the bow in the front there is a control station, which means that the vessel can either be controlled completely from here or from the bridge. The advantage of having an operator in the bow is that he can see everything that is in front of him, so that he can direct the vessel right into the thickest part of the slick. This vantage point also makes it easier to avoid large pieces of debris.

In Figures 17 and 18 the vessel is recovering white foam chips. The skimmer is very maneuverable. This charasteristic is valuable if a spill becomes dispersed into smaller slicks.

Figures 19 through 23 show the skimmer being equipped with floating booms to provide a very wide sweep. These are 150-foot lengths of boom that are deployed in a V configuration. It is apparent that the Coast Guard provided very active and helpful participation in this demonstration. The Captain of the Port of Philadelphia, John Kirkland, was most cooperative and enthusiastic during the exercises. To operate three vessels in such a mode requires coordination, communication, and good seamanship, but it was demonstrated that it could be done expeditiously.

In Figure 24, the foam chips are being released.

Figures 25, 26, and 27 are successive shots of the chips: being gathered by the V-booms; approaching the mouth of the skimmer; and finally being recovered. The results attained in this demonstration are exactly what we would expect under similar weather conditions in a real oil recovery operation.

Figure 28 is the collection well. This was shown in the drawing on Figure 14. It is here that debris and oil are captured at the rear of the belt. Figures 29, 30, and 31 show a clamshell bucket, which is used to remove debris. The recovered oil is pumped away. There are 10,000 gallons of on-board storage on this vessel. If larger amounts are involved, it can be pumped into a barge or a tanker.

The discussion to this point has focused on containment and removal. Another technique that seems to be coming into its own, although it is not fully accepted yet, is the use of dispersants for cleaning up spills. Now, the use of dispersants is restricted under the National Contingency Plan. Permission must be obtained from the regional response team. That means it is illegal to use dispersants on a spill unless permission is granted by the regional response team.

There are twelve dispersants that have been reviewed by EPA and found to be acceptable for use in spill situations, but remember that permission must be obtained on a case-by-case basis. There are two kinds of dispersants. One type shown in Figure 32 requires mixing energy to make it effective. The dispersant is applied to a spill, then some mixing energy, and the oil forms into droplets. They gradually dissipate and become more readily accessible for biodegradation.

The other type of dispersant shown in Figure 33 is the self-mixing kind. It is applied in much the same way, but the physical chemistry forces that are at work cause the dispersant to migrate through the oil to the water. In doing so, as it moves through the slick, it goes right through the boundary layer and it takes some oil with it and forms droplets, thereby requiring no mixing energy.

Figure 37 is a picture taken in British Columbia, of a demonstration sponsored by Exxon using a Conair multi-engine aircraft to spray dispersants. The purpose of this work was to demonstrate the feasibility of the concept in general and more specifically the ability to control droplet size, application rate, and efficient application of the dispersant on the slick. A number of demonstrations conducted since then by the American Petroleum Institute, in conjunction with EPA, have shown that this is a viable technique for applying dispersant to an oil spill.

Figure 38 shows a multi-engine craft spraying dyed dispersant. The dispersant was dyed to facilitate observation of the spray pattern. This test was conducted in southern California. API has conducted similar tests in the Atlantic, off New Jersey. This is recent work. These findings and results will be published in technical papers and also through publications of the American Petroleum Institute.

Figure 39 is a dispersed slick. The photograph was taken probably not more than ten minutes after the application of dispersant. A plume form which has a smoky appearance. The tiny oil-dispersant droplets are dissipating throughout the water column. In this state the oil is more amenable to biodegradation.

Figure 40 shows the sampling pattern in one of the tests and Figure 41 shows the results of sampling at station 8 of time with depth. The results indicate that the oil was quickly and efficiently dispersed. Consideration at the surface are initially what might be expected; however, after a rather short period of time there are no high concentrations anywhere, either on the surface or at varying depths. We believe that dispersants can be a useful tool in mitigating the harmful effects of oil spills and should be given an increasingly greater role in response actions.

Mr. Kazmierczak is Coordinator of Energy Conservation and Oil Spills Control Technology for the Sun Company, Inc., in Philadelphia, Pennsylvania. He received a Bachelor of Chemical Engineering degree from Villanova University in 1955 and a Master of Science in Environmental Engineering from Drexel University in 1973. Mr. Kazmierczak deals with environmental problems raised by petroleum industry activities and special problems involving synthetic fuel development. He has written and lectured on Outer Continental Shelf activities and related environmental problems (especially oil spills and energy conservation).

Figure 1   Boom storage container

Figure 2   Abrasion pad

Figure 3   Initiation of boom deployment

Figure 4   Partly extended boom.   Note outline
of internal coil

Figure 5   Left section of boom extended
             Right section partly extended

Figure 6   Fully extended boom showing wave conformance
             characteristics

Figure 7   Another view showing wave conformance

Figure 8   Larger size Kepner boom being tested

Figure 9   Another view of larger size Kepner boom

Figure 10   STOPS portable pumping system

Figure 11   Another view of STOPS portable pumping system

Figure 12   Cushion on fender for use with vessels
             alongside each other

Figure 13   Goodyear boom being deployed over a roller

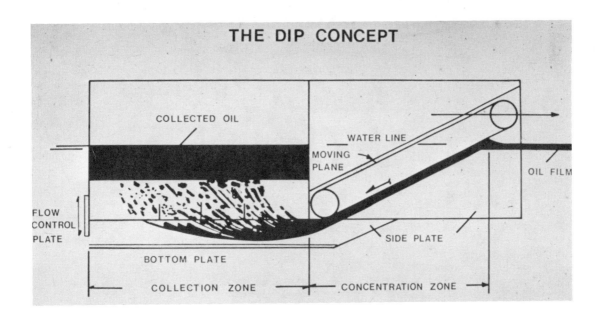

Figure 14

Figure 15   Full view of Delbay skimmer

Figure 16   View of sweeps and opening to collection system

Figure 17   Plastic foam chips approaching opening

Figure 18   Same as Figure 17

Figure 19   Beginning of boom deployment for V-configuration

Figure 20   Continuation of boom deployment

Figure 21   Continuation of boom deployment

Figure 22   Deployment of boom in the V-configuration

Figure 23   Same as Figure 22

Figure 24   Plastic foam chips being released for
recovery test

Figure 25   Skimmer approaching floating foam chips

Figure 26   Chips inside V-boom just prior to recovery

Figure 27   Immediately prior to recovery

Figure 28   Recovery well

Figure 29   Clamshell debris bucket

Figure 30   Same as Figure 29

Figure 31   Recovered chips being removed from well
by clamshell

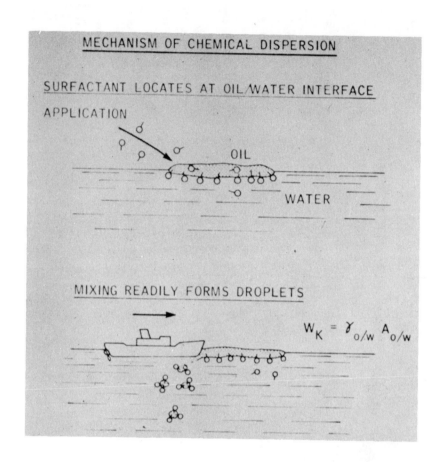

Figure 32

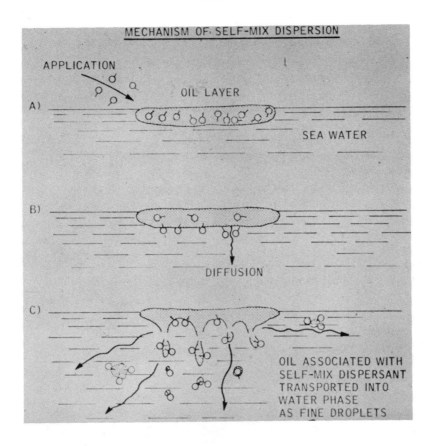

Figure 33

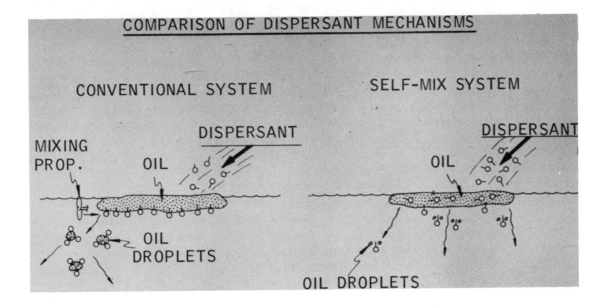

Figure 34

Figure 35   Dispersant spraying using hose nozzle

Figure 36   Dispersant spraying through booms
             followed by breaker-board mixing

Figure 37   Dispersant spraying by multi-engine
             aircraft in Canada

Figure 38   Dispersant spraying by multi-engine aircraft--
West Coast, U.S.

Figure 39   Dispersed oil plume just after dispersant
application

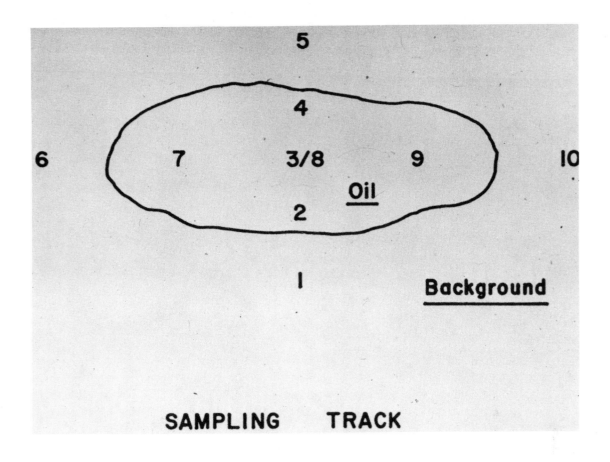

Figure 40   Sampling track of API-EPA dispersant test

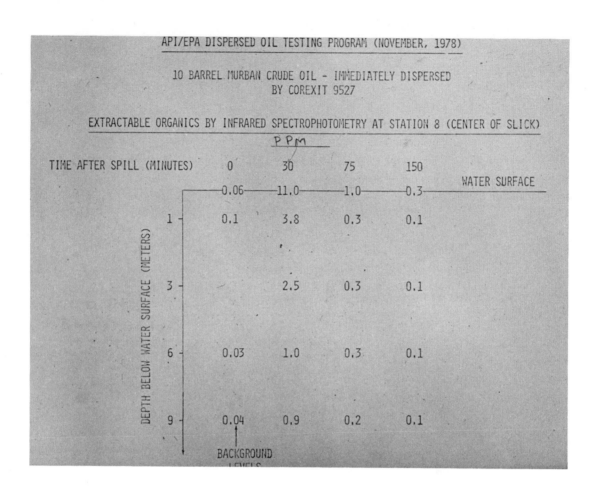

Figure 41   Results of API-EPA dispersant test

DISCUSSION

DR. RICKARD: This concludes the prepared part of our session today, and we can entertain questions now both for the afternoon and the morning speakers. I would like to ask you to please identify yourself and your organization as you ask a question.

MR. KRAHL: I am Dick Krahl of the U.S. Geological Survey. I would like to ask a couple questions. This morning Jim Albers talked about two rather sophisticated devices for deep water exploration, including the recoilless riser system and multiplex blowout control system. I would like to know at what water depths these systems should be considered for use.

MR. GEER: I will try to answer your question for Jim. As far as a straight hydraulic control system is concerned and as far as time response for preventer closure, you are probably limited, to a certain degree, in water deeper than about 1,000 feet. As you start to look for additional feedback in terms of monitoring positions of rams, you are limited by the number of signals that can be effectively handled with a straight hydraulic system. Then you may require more sophisticated systems such as electrohydraulic systems. That is where you find yourself needing a multiplex system because there are economic as well as physical limits, as most of you know who have used instrumentation in deep water. You are limited by the number of conductors you would want to handle in an umbilical cord. What you are trying to do is trade off between the requirements and the signal response and how many signals you can crowd over a single carrier, and by limiting the number of conductors, and yet still have redundancy to compensate if a conductor has failed. An umbilical cord, or course, costs a lot of money. You then try to limit your signal patch requirements. But if you have a number of spares in case one of them goes haywire, you can just change the connections. I would say, generally speaking, when you get beyond 1,000 feet of water, you will probably find yourself going towards a multiplex system. There are varying degrees of this. It depends on who the contractors and operators are and where you are operating, but I think in general this is probably Sedco's approach. Now, as far as the recoilless system is concerned, I do not know exactly at what depth SEDCO has had experience but, as you probably know, there is a tremendous amount of energy stored in a marine riser tensioning system under normal operating conditions. Here again, it just depends upon how much of the marine riser tension is based upon buoyancy built into the riser system and how much is actually handled from physical tensile loading at the surface through the tensioning system. It is a case of trying to balance the number of tensioners on the surface and the physical limitations of the vessel with the depth capability. If you are in 2,000 to 4,000 feet of water, that is one thing. If you are in 4,000 to 6,000 and maybe trying to stretch to 8,000 feet, as we were talking about today, that is another thing. But here again, it is a balancing of the space on the vessel in conjunction with the diameter of the riser, the amount of buoyancy on the riser, the diameter of that buoyancy material on the riser, and the amount of tension capacity of a Rucker unit, say, or a Vetco unit. I would say in this case probably 2,000 feet deep is when you start getting into a more sophisticated system because of the physical dimensions of the riser itself (length and diameter) and the physical limitations on the rig to accommodate that additional riser length and the greater surface tensioning requirements. A considerable number of wells have been drilled in less than 2,500 feet of water without a recoilless system, but as the water depth approaches 5,000 feet it is expected that a recoilless riser tensioning system will be required. As I said earlier this morning, there are so many variables that must be considered that there is no magical crossover point where one stops one type of operation and starts another. It is a matter of degree. If you've got a piece of equipment, let's say a drilling rig, that has certain inate capabilities, and you are trying to extend those capabilities, you do one thing in a tradeoff to accommodate the additional depth. If you are building a new vessel from scratch, you can spend a little more time, a little more money, perhaps build a little more capacity into it to accommodate those things. So it is a variation.

MR. KRAHL: Thank you. Another point. I would like to ask Hans Jahns if there are deep water areas around the United States that would lend themselves to construction of a Condeep or Seatank type structure, or in what areas could those be built?

MR. JAHNS: You are talking about sea ice areas?

MR. KRAHL: No, I am talking about the deep water areas close to land that would lend themselves to such construction, like the Norwegian fjords that were used for construction sites.

MR. JAHNS: I do not really have much to say in answer to that. Of course, there are the correct water depths somewhere along the continental shelf or the continental slope where a structure of that kind could be used.

MR. DUNN: We did a study for the Gulf of Alaska a few years ago, and we found two suitable sites on the West Coast, one in Alaska, one in Puget Sound, and then there is a third possibility in Canada near Vancouver, British Columbia. We would have had to modify the concrete structure, but we felt that we could build them, in one of these locations, for about 500 feet of water.

MR. KRAHL: A question was asked this morning concerning removal of platforms. Would it not be true that when you consider the economics of a project, the costs of removing the platform would be cranked into the total equation, and in so doing, you would have had to decide on some sort of removal procedures?

MR. DUNN: Yes, this is correct. We estimated the cost, but I don't remember just what we estimated for Cognac. We normally use about 1/3 to 1/2 of the cost of the original installation cost for salvage. Keep in mind, there is money to be spent 30 years down the road, and it is really not worth very much on a present day value economic basis.

MR. KRAHL: The point that I am making is that if, indeed, you do an economic analysis of this, there has to be some sort of engineering analysis done as to the method of removal for applying the economics.

MR. DUNN: There was no detailed engineering analysis done in the case of Cognac. We knew it could be done, so we just used a rough estimate.

MR. KRAHL: I know the question has come up in the past as to removal of the structures in the upper Cook Inlet, and maybe an analysis has not been done for the case.

MR. DUNN: Salvage of those platforms would not be a difficult thing to do.

MR. KRAHL: One other point that would be addressed to Jim Ray concerning the mud. He addressed many of the costs, and we recognize the costs that are going to be levied or that would be levied due to the use of, for instance, barge muds. He also mentioned some studies that have been done but did not get into the results of any of these studies. I am hoping that the paper that he is preparing, or the proceedings, will indicate the results of these studies from which one might draw some analysis as to the economic tradeoffs versus the environmental tradeoffs that might be made.

MR. GEER: In his brief presentation and also in the documentation of his brief presentation, he did not attempt to go into the economic tradeoffs as such. It is my understanding that those numbers were previously presented in testimony before certain congressional hearings in conjunction with the OCS sales on the East Coast, both in the mid-Atlantic area as well as the one in preparation for the Georges Bank area. It is, further, my understanding that there will be more a quantitative, factual presentation of these kinds of data at the symposium on Research on Environmental Fate and Effects of Drilling Fluids and Cuttings to be held in Orlando,

Florida, on January 21-24, 1980. As you can appreciate, we had a very limited time to try to cover this subject in more detail today. As far as a specific tradeoff is concerned, it is my understanding that the Offshore Operators Committee (OOC) did present or provide Mr. Beller of the Environmental Protection Agency quite a compilation of data about a year ago that would back up the kinds of numbers that were mentioned here today in a rather cursory way.

MR. MASON: I might add a couple of points.
The Discoverer Seven Seas is equipped with a riser recoil system, and I cannot say whether it was operating in the well that was just finished in 4,800 feet of water. It was operating off Spain in about 3,900 feet, as I remember. As far as as the control systems are concerned, I do not know what the water depth limit is going to be on these electrohydraulic systems, but it is my opinion that because of electrical connectors at the present time 3,000 to 4,000 feet is probably the limit. Right now the BOP control systems are designed so that they can be retrieved without recovering the BOP stack. It is my opinion that there will be a tradeoff at some point, at which time you can decide against retrieving the control modules. Instead, you will pull the riser if they malfunction. A redundant system is on these. They surely would not both fail at the same time. So I think the electrohydraulic system can probably be used at just about any desired depth. Somewhere the decision would be made not to pull the control modules without pulling the BOP stack.

MR. KRAHL: I think one of the points I was looking for was the lower depth limit.

MR. MASON: The upper tradeoff. Do you mean the shallower water or the deep--

MR. KRAHL: The shallower waters. That is what Ron was addressing.

MR. ELA: My name is Deke Ela. Would Mr. Hughes care to comment on the sea state limitations of the diving systems that he discussed?

MR. HUGHES: Obviously, the limitation is strictly in regard to launch and recovery of each of those systems. I wish I could give you some accurate numbers. We have launched and recovered various forms of diving systems in wave heights in exceeding 20 feet. There are a number of provisions you may make to improve the rough water launch and recovery capability. In fact, in most areas where we have supported exploration and production, we have not had to design an optimum system in order to supply the service needed. I do not think anyone has yet been faced with an environmental situation that stretched our design capability to the limits. However, we have some ideas that we have not used yet. When you launch and recover a system like that in relatively high seas, it is best to use a guidance rail system which essentially gives you a hard fast grip on the system until it is below the interface as well out of most of the wave action. That is a simply a matter of designing launch and recovery carriages and cage mechanisms. There are also some ideas about launch and recovery through the hull itself, particularly on semi-submersible rigs, which could improve launch and recovery character-istics in rough weather but, again, we have not really been stretched to the limits yet.

MR. SMEDLEY: Patrick Smedley, Lloyd's Register of Shipping, London, England. One issue that has been glossed over today on this program is repair. We have heard something about repair of pipe-lines, and I accept that today it does not pose too serious a problem. We know that in underwater habitats we can make welds or at least they can be made as good or better than on land. But the issue that has been rather neglected is what we do about the larger structures, whether steel or concrete. Now, to me, this is also a national problem because the longer installations are shut down the worse the effect can be on the national economy. In the United Kingdom, we are at least trying to encourage the government to take an interest in this issue. Every time there is a need to repair a structure, not only is it extremely expensive, it involves quite a considerable amount of thought as to how it can be done, and the techniques that can be used. I wondered what the views were of the others who have been concerned with this aspect.

MR. DUNN: You are correct. It is a potentially serious problem. Several platforms in the Gulf of Mexico and in Cook Inlet, as a matter of fact, have been repaired using various techniques. In the Gulf of Mexico, for example, we have used dry welding to repair fractures or cracks. Wet welding has been used also but it does not yet quite match the quality of dry welding. In Cook Inlet, we have replaced complete members. It is most assuredly not cheap, and it is difficult. However it can be done. I did not mention it this morning, but I think building a good deal of redundancy into struc-tures is the best cure. If you have lost one member, you still do not have a seriously damaged platform. I think it is most important for the engineer to keep that in mind. The other thing is to consider just what has caused these problems. Large ship-platform collisions--I do not really know what you can do about those. It is going to happen. It has happened to major platforms about seven or eight times in the past 20 years, with complete loss of the plat-forms. We have also had several relatively minor collisions. One of them took place, as a matter of fact, just last week. People drop things overboard. That used to be the bane of our platform repair problem. I think it is most important to stop such carelessness. We have been quite successful in minimizing this in the past few years. I really do not know much about repair of concrete platforms nor do I know whether any repairs have been made on such platforms. Mike, do you have anything to add on wet welding and dry welding?

MR. HUGHES: Since we spend much of our efforts in inspecting and repairing offshore facilities, we have become familiar with the nature of damage and cause of damage and the types of repairs that are possible. It becomes quickly evident to us that most of the damage to offshore structures occurs near the surface, such as collisions by ships and barges, and is generally restricted to parts of the platform in shallower water. I cannot recall a single major underwater repair required for an offshore platform near the bottom of the platform. There has been damage through dropped piling and other objects, as Pat pointed out, which was repaired, but in no case that I can recall were those repairs considered to be essential to the integrity of the platforms. If there is a failure near the bottom, it is generally a catastrophic failure, and the whole platform goes. Then you do not have to worry about a repair; you have a salvage operation. I do not mean to minimize the difficulty of repairing in deep water, but simply to point out that our experience has been that the vast majority of repairs do take place on the parts of the platforms that are in shallow waters.

MR. DUNN: I might mention one other thing. We have done some repairs on some of our older platforms because of failure of the cathodic protection system. Here again, I think it is extremely important to ensure that a reliable cathodic protection system has been installed initially, because that is one of the most important design factors.

MR. BLAKE: Ken Blake from the USGS. I would like to ask Hans Jahns about the ice islands and what kind of subsea stack and riser system is used for the drilling, and how would he keep the ice island from moving if we ever did get to the point where they could use them all year round?

MR. JAHNS: So far only one ice island has been used to drill a well, and in that operation there was no subsea preventer system. There was a big caisson, I think, 72 inches in diameter and 2 inches in wall thickness that was driven into the sea floor through the island, which would have provided some protection for the well inside should the island have shifted somewhat. But as I mentioned, the island itself was protected by an 11-foot wide moat. If we talk about a two-year ice island operation, presumably we would have an ice island that is in itself large enough and massive enough to resist iceloads, just as a gravel island does. Our objective with the ice island that Exxon built was that it would have about the same lateral resistance capability as a typical gravel island, so it would on that basis provide the same resistance, the same safety, as a gravel island would. Of course, ice islands could also be used in connection with sub-mudline BOP's, if that is necessary. This would be similar to drilling operations in Canada's High Arctic floating ice platforms. They are equipped with seafloor BOP systems. Those operations are in water 400 feet to 1,000 feet.

MR. HEALEY: Lieutenant Ted Healey, Coast Guard. I have a question for Mr. Hughes. The multinational nature of the diving industry may give you a few problems. How is the industry going about coping with standards of different nations?

MR. HUGHES: I suppose the most serious effort for international standards would be through the Inter-Governmental Maritime Consultative Organization, and of course we are maintaining contact through the Inter-Governmental Maritime Consultative Organization representatives with the activities. Since this country has pioneered much of the underwater activity around the world, we have had the opportunity to have some input to all of the countries regarding their underwater safety requirements. I spent quite a lot of time with various agencies in the United Kingdom in the formulation of underwater safety rules. There is not a great deal of disparity between the countries. There is some, and it is a problem because a system that is prefectly adequate in one country may be deemed inadequate in another, and we are following that primarily through IMCO.

MR. LEES: John Lees with the U.S. Geological Survey. I would like to inquire about what procedures have been developed or are in the process of being developed for operating production equipment on places where you have unstable slopes and possible bottom sediment problems.

MR. GEER: Do you refer to bottom-founded structures, or are you speaking with regard to subsea completions or combinations thereof?

MR. LEES: Subsea completions, mainly.

MR. GEER: We always look at what the foundation requirements are and then build the foundation pile to accommodate whatever the soil requirements dictate. Therefore, if you have an overlying area of a given depth, of which it is suspected that the material is likely to move, then you take into full consideration what your pile shear loads will be in bending--that is the moments that they will have to withstand to accommodate whatever is expected to occur during the life of the production operation. This is taken into account not only with subsea well completions in the foundation piles, but it is also looked into and is a design criteria for the piling for any of our offshore structures. A case in point is Cognac and some other structures. That technology has advanced considerably since the late 1960's when we lost one of our structures due to a mud slide in the Gulf of Mexico.

MR. DUNN: There really is no difference in principle or in fact from the manner in which we design slide resistant platforms. There are now about eight slide resistant platforms in service in the Gulf of Mexico.

CAPT. CRONK: I am Peter Cronk, U.S. Coast Guard. I have a question concerning tension leg platforms. What standards are used or may be most applicable to the design of the hull?

MR. GEER: We have done a good bit of work on this. I would assume that the same standards that are used for any mobile floating drilling unit such as semi-submersibles would be considered. At least that is the initial criteria that we consider in the economics of a given venture, taking into account such alternatives as guyed towers, articulated columns, or subsea completions. We try to look at all of these, but as far as the criteria or standards which we would apply to a tension leg platform, it would be in the same mode as a mobile drilling rig such as a semi-submersible.

SUMMARY

James A. Rickard

I want to thank the speakers very much. They spent a lot of their time developing special presentations for this occasion. These are not canned talks you heard today. We asked speakers to cover specific topics and in many cases they had to start from scratch.

I want to thank the Marine Board for its support of this conference and, most of all, I want to thank the attendees. It took a lot of time and money to put this symposium on, but it is worthwhile if it will help you in the very important work you do.

I was talking with one of the USGS people at lunch and was impressed, as he was, by the number of important decisions that are made in our government and with the big dollars and the big stakes that our public has in those decisions. So we really feel this has been worthwhile if we have helped you somehow in the work you do to serve the public.

It is customary at the end to summarize in brief fashion what every speaker said, but I am not going to do that today, except in a very negative way. What I am going to do is review the notes I made as to what the speakers either did not say they could do, or said they could not do--in other words, what they indicated are the current limits to the technology.

We have heard them say that they cannot actually drill an exploratory well in waters deeper than 6,000 to 8,000 feet, although the technology is available. They cannot drill one in water that deep if the current is greater than 4 knots or if the waves average more than 15 feet high. So if you propose a lease sale but the tracts do not meet those qualifications, the petroleum industry currently cannot develop them.

You have heard the speakers say that they cannot build a fixed bed platform that is economically competitive in water deeper than about 1,200 to 1,500 feet and compliant platforms are not competitive after 2,500 feet. Furthermore, right now the technology does not exist to build a subsea template or subsea well with the facilities that go with them, deeper than about 3,900 or 4,000 feet of water.

Hans Jahns says he could not build a gravel island in the Beaufort Sea in waters deeper than about 60 feet and that platform technology for deeper arctic water (up to 100 feet) would not be available before the late 1980's. He did say he had preliminary designs for 200 feet in the Canadian Beaufort Sea, but that is as far as he has progressed.

Operators cannot lay pipelines of reasonable size in waters deeper than 3,000 feet today. The biggest pipe that can be laid is 36 inches in diameter. The size of the diameter is limited by the mills' incentive to make the pipe, though, not by any inherent inadequacy in the pipelaying barge.

No one has talked about an offshore floating production system in water depths more than about 5,000 feet, so if you have a tract that is 6,000 feet deep, we cannot build a floating platform there today.

I guess the star of our show as far as capabilities are concerned is Mike Hughes, who says he can put a deep water supporting system under water anywhere we can operate.

Thank you very much. These proceedings are closed.

Dr. Rickard is a member of the Marine Board and is Manager, Planning, Exxon Production Research Company in Houston, Texas. He attended Iowa State and Texas A & I University (BS, MS, 1948) and the University of Texas (Ph.D, Physics, 1953). He joined the former Humble Oil and Refining Company as a research engineer in 1953 and worked in various petroleum production and offshore research areas. In 1968, he became General Manager at Exxon Production Research Company. He joined Exxon Corporation in New York in 1971, serving in executive positions in the Environmental Conservation, Science and Technology, and Producing Departments.

PARTICIPANTS

J. C. Albers
SEDCO, Inc.
Dallas, TX

Henry S. Anderson
Office of Marine Sciences
  & Technology Affairs
Office of Oceans & Fisheries
  Affairs
U.S. Department of State
Washington, DC

Willis C. Barnes
ORI, Inc.
Silver Spring, MD

Robert H. Barr
U.S. Coast Guard
Washington, DC

Thomas J. Barrett
U.S. Coast Guard
Washington, DC

William Beller
U.S. Environmental Protection
  Agency
Washington, DC

Eric Bender
Sea Technology Magazine
Arlington, VA

Michael Bender
Virginia Institute of Marine
  Science
Gloucester Point, VA

Kenneth Blake
U.S. Geological Survey
Washington, DC

Reed M. Bohne
National Oceanic & Atmoshperic
  Administration
Washington, DC

Helen S. Bolton
U.S. Environmental Protection
  Agency
Washington, DC

Dennis J. Cashman
U.S. Coast Guard
Washington, DC

Thomas Charlton
U.S. Environmental Protection
  Agency
Washington, DC

Sarah Chasis
Natural Resources Defense
  Council, Inc.
New York, NY

D. M. Coleman
Mobil Research & Development
  Corp.
New York, NY

Clare J. Colman
Mobil Research & Development
  Corp.
New York, NY

Ulysses Cotton
U.S. Geological Survey
Reston, VA

Peter J. Cronk
U.S. Coast Guard
Washington, DC

Bud Danenberger
U.S. Geological Survey
Reston, VA

James T. Dean
Global Marine Development,
  Inc.
Newport Beach, CA

J. E. DeCateret
U.S. Coast Guard
Washington, DC

Kent Dirlam
Bureau of Land Management
Washington, DC

Norman Doelling
MIT Sea Grant Program
Cambridge, MA

David Duke
U.S. Department of Energy
Washington, DC

F. P. Dunn
Shell Oil Company
Houston, TX

N. Terence Edgar
U.S. Geological Survey
Reston, VA

Charles N. Ehler
National Oceanic & Atmospheric
  Administration
Washington, DC

* Phillip Eisenberg
  Hydronautics, Inc.
  Washington, DC

Deke Ela
Arnold, MD

Bob Ensminger
Conservation Division,
  Easter Region
U.S. Geological Survey
Washington, DC

* Davis Ford
  Engineering Science Company
  Austin, TX

John Freund
Naval Sea Systems Command
Washington, DC

* William S. Gaither
  University of Delaware
  Newark, DE

Richard Giangerelli
U.S. Geological Survey
Reston, VA

* Ronald L. Geer
  Shell Oil Company
  Houston, TX

* Ben C. Gerwick, Jr.
  University of California
  Berkeley, CA

Richard E. Haas
U.S. Coast Guard
Washington, DC

Cheryl Hawkins
U.S. Environmental Protection
  Agency
Washington, DC

---

*   Member of the Marine Board

Timothy C. Healey
U.S. Coast Guard
Washington, DC

Donald C. Hientze
National Ocean Industries
  Assn.
Washington, DC

David Holton
U.S. Department of State
Washington, DC

John C. Houghton
Office of Science & Technology
  Policy
Washington, DC

Michael Hughes
Oceaneering International
Houston, TX

Hans O. Jahns
Exxon Production Research
  Company
Houston, TX

Raymond A. Karam
U.S. Department of the
  Interior
Washington, DC

Don E. Kash
U.S. Geological Survey
Reston, VA

Leon J. Kazmierczak
The Sun Company
Philadelphia, PA

Charles H. King, Jr.
U.S. Coast Guard
Washington, DC

George L. Kinter
National Oceanic &
  Atmospheric Administration
Washington, DC

Joseph D. Klimas
U.S. Coast Guard
Washington, DC

Richard B. Krahl
U.S. Geological Survey
Reston, VA

John Lees
U.S. Geological Survey
Washington, DC

J. T. Leigh
U.S. Coast Guard
Washington, DC

Cesar De Leon
U.S. Department of
  Transportation
Washington, DC

William Linder
Petro-Marine Engineering, Inc.
Gretna, LA

J. C. Lochridge
Brown & Root, Inc.
Houston, TX

B. L. Marcum
Mobil Oil Corporation
New York, NY

J. Preston Mason
Seaflo Systems, Inc.
Houston, TX

Stephen J. Masse
U.S. Coast Guard
Washington, DC

George F. Mechlin
Westinghouse Electric Corp.
Pittsburgh, PA

*Leonard Meeker
 Center for Law and Social
   Policy
 Washington, DC

H. William Menard
U.S. Geological Survey
Reston, VA

O. E. Van Meter, Jr.
Mobil Exploration and Producing
   Services, Inc.
Dallas, TX

Rufus Morison
U.S. Environmental Protection
   Agency
Washington, DC

*J. Robert Moore
 University of Texas at Austin
 Austin, TX

John Moroney
Tulane University
New Orleans, LA

W. Michael Mulcahy
Sea Technology Magazine
Arlington, VA

Hyla S. Napadensky
Illinois Institute of Technology
Chicago, IL

William M. Nicholson
National Oceanic & Atmospheric
   Adminstration
Rockville, MD

Lawrence J. Nivert
U.S. Coast Guard
Washington, DC

*Myron H. Nordquist
 Nossaman, Krueger & Marsh
 Washington, DC

Elliot A. Norse
U.S. Council on Environmental
   Quality
Washington, DC

Virginia Fox Norse
U.S. Environmental Protection
   Agency
Washington, DC

Jose Notario
U.S. Department of the
   Interior
New York, NY

Harley D. Nygren
National Oceanic & Atmospheric
   Administration
Rockville, MD

Chris Oynes
U.S. Department of the
   Interior
Washington, DC

Dean Parson
National Marine Fisheries
   Service
Washington, DC

---

* Member of the Marine Board

Charles Perrott
U.S. Geological Survey
Reston, VA

T. J. Polgar
U.S. Coast Guard
Washington, DC

Ronald Prehoda
U.S. Geological Survey
Reston, VA

Alan H. Purdy
National Institute for
  Occupational Safety and
  Health
Rockville, MD

James P. Ray
Shell Oil Company
Houston, TX

Paul Purser
Houston, TX

Jerry Richard
U.S. Geological Survey
Reston, VA

Norman L. Richards
U.S. Environmental Protection
  Agency
Sabine Island
Gulf Breeze, Florida

*James A. Rickard
Exxon Production Research Company
Houston, TX

Robert L. Rioux
U.S. Geological Survey
Reston, VA

T. H. Robinson
U.S. Coast Guard
Washington, DC

Michael P. Rolman
U.S. Coast Guard
Washington, DC

James Rucker
National Oceanic &
  Atmospheric Administration
Rockville, MD

Richard Rumke
Ship Research Committee
National Research Council
Washington, DC

James A. Sanial
U.S. Coast Guard
Washington, DC

*Willard F. Searle, Jr.
Searle Consultants, Inc.
Alexandria, VA

Wilbur G. Sherwood
National Science Foundation
Washington, DC

O. J. Shirley
Shell Oil Company
New Orleans, LA

Michael Silka
U.S. Coast Guard
Washington, DC

Ben Silverstein
Office of Technology Assessment
U.S. Congress
Washington, DC

Ed Simonis
U.S. Geological Survey
Washington, DC

Phillip S. Sizer
Otis Engineering Corp.
Dallas, TX

---

* Member of the Marine Board

George P. Smedley
Lloyds Register of Shipping
London, England

Harold Smith
U.S. Geological Survey
Washington, DC

Jeffrey B. Smith
U.S. Department of Energy
Washington, DC

J. Edward Snyder, Jr.
U.S. Navy (Ret.)
McLean, VA

George Sorkin
U.S. Navy
Washington, DC

James E. Steele
Naval Architect
Quackertown, PA

Sherry Steffel
House OCS Committee
Washington, DC

Wayne Stevens
Bureau of Land Management
U.S. Department of the
  Interior
Washington, DC

Newell Stiles
U.S. Geological Survey
Washington, DC

Robert H. Stockman
National Oceanic &
  Atmospheric Administration
Washington, DC

Paul G. Teleki
U.S. Geological Survey
Reston, VA

*James G. Wenzel
 Lockheed Missiles &
   Space Company, Inc.
 Sunnyvale, CA

*Robert L. Wiegel
 University of California
 Berkeley, CA

Frank L. Whipple
U.S. Coast Guard
Washington, DC

Stearns Whitney
U.S. Coast Guard
Washington, DC

Peter E. Wilkniss
Ocean Sediment Coring Program
National Science Foundation
Washington, DC

Robert S. Winoker
Naval Oceanographer Division
Washington, DC

Lawrence R. Zeitlin
Lakeview Research, Inc.
Peekskill, NY

---

* Member of the Marine Board

Date Due